Advance Praise

Gardening [illegible]

I recommend *Gardening When It* [illegible] he beginning gardener. It includes all [illegible] ake, prepare, and tend a garden, make compost, choose varieties and seeds, and to grow just about everything. I also recommend it to all experienced gardeners. It is guaranteed to change the way you view and do gardening. G*ardening When it Counts* is a magnificent synthesis of garden science, original garden research, and agricultural history.

— Carol Deppe, Ph.D., author of *Breed Your Own Vegetable Varieties: The Gardener's and Farmer's Guide to Plant Breeding and Seed Saving.*

In *Gardening When it Counts,* Steve Solomon prepares the ground by encouraging us to embrace the organic revolution by growing more food with fewer imported resources and more ingenuity. He provides a hands-on account of amendment-centered gardening, using a wide variety of sources, and exhorts us to save seeds of kinds and varieties that we like to eat, that do well for us, and that may be dropped from current commercial seed inventories.

— Alan M. Kapuler, Ph.D., President, Peace Seeds, and former Research Director and cofounder of Seeds of Change

Steve Solomon's book is delightfully informative and abundantly rich with humor and grandfatherly wisdom. A must-read for anyone wanting a feast off the land of their own making ....

— Elaine Smitha, host of "Evolving Ideas" radio and telvision, and author of *If You Make The Rules, How Come You're Not Boss?*

Books for Wiser Living from *Mother Earth News*

Today, more than ever before, our society is seeking ways to live more conscientiously. To help bring you the very best inspiration and information about greener, more sustainable lifestyles, New Society Publishers has joined forces with *Mother Earth News*. For more than 30 years, *Mother Earth News* has been North America's "Original Guide to Living Wisely," creating books and magazines for people with a passion for self-reliance and a desire to live in harmony with nature. Across the countryside and in our cities, New Society Publishers and *Mother Earth News* are leading the way to a wiser, more sustainable world.

Steve Solomon

gardening
when it counts

GROWING FOOD *in* HARD TIMES

NEW SOCIETY PUBLISHERS

Cataloging in Publication Data:
A catalog record for this publication is available from the National Library of Canada.

Cover design by Diane McIntosh. Photo: Getty Images/Photodisc.
All illustrations by Muriel Chen unless otherwise credited.

Printed in Canada.
Eighth printing April 2009.

Paperback ISBN: 978-0-86571-553-0

To order directly from the publishers, please call toll-free (North America) 1-800-567-6772, or order online at www.newsociety.com

Any other inquiries can be directed by mail to:
New Society Publishers
P.O. Box 189, Gabriola Island, BC V0R 1X0, Canada
1-800-567-6772

New Society Publishers' mission is to publish books that contribute in fundamental ways to building an ecologically sustainable and just society, and to do so with the least possible impact on the environment, in a manner that models this vision. We are committed to doing this not just through education, but through action. We are acting on our commitment to the world's remaining ancient forests by phasing out our paper supply from ancient forests worldwide. This book is one step toward ending global deforestation and climate change. It is printed on acid-free paper that is **100% old growth forest-free** (100% post-consumer recycled), processed chlorine free, and printed with vegetable-based, low-VOC inks. For further information, or to browse our full list of books and purchase securely, visit our website at: www.newsociety.com

NEW SOCIETY PUBLISHERS www.newsociety.com

Dedication

Since the time I was sent to elementary school, my feet have marched to the beat of a different drummer than Everybody Else's. The difference caused me much grief as a child; much success as an adult.

This book is for Mr. and Ms. Everybody Else, who are well known and highly respected authorities on most everything, including vegetable gardening. Most people look to Everybody Else for guidance before making any decision. It has long been my experience that Everybody Else is often wrong and needs the information in this book.

Contents

CHAPTER 1

Introduction

During the 1970s, inflation and unemployment were high. In such lean years, many people grow substantial backyard veggie gardens. I was a young man who did that.

Good times returned in the 1980s and continued into the first half of the first decade of the new millennium — fat years. I was there. In easy times people go to restaurants and take summer vacations; not me, I continued gardening.

Now I am 63 years old, still flexible enough to touch my toes (on a good day), still able to put in a hard morning's work, still growing the majority of what we eat in our household, still doing it 12 months a year. These days I feel fortunate to have retired to one of the world's most remote places, Tasmania, a temperate South Pacific island with a climate that is a lot like Oregon's. From here I can enjoy a slight sense of detachment as I watch how the planet is going. But Tasmania is not self-sufficient, so I am not nearly as detached as I wish I could be about the hard times I foresee coming. I have the feeling that I should share some gardening knowledge I've accumulated with those who are probably soon going to need it, which is why I wrote this book.

During the fat years, an unfortunate change happened in veggie gardening. Books and magazine articles promoting traditional homestead and backyard methods — growing well-separated plants in rows far enough apart that you could walk between them — disappeared. Row gardening was universally denounced as a waste of space, inefficient with water, and low-producing. Densely packed, deeply dug, super-fertile, massively irrigated, raised-bed systems became fashionable. As I write this book in 2005, intensive gardening still reigns.

When I started suburban backyard food gardening, John Jeavons was just starting to write about intensive gardening, and that was the method I used. Five years later I became a back-to-the-lander and continued to garden intensively even though I had a five-acre (two-hectare) homestead and could have spread my plots out as widely as I wished. In 1979 I created Territorial Seed Company, a homestead-based mail-order vegetable-garden seed business, and by 1984 I had written three gardening books recommending intensive methods.

During the 1980s, when intensive had become standard practice, several things came together to teach me it was not the best way. Because I was running a seed company, I had to do variety trials. What are variety trials, you ask? Well, an honest seed business does not sell just any old variety of seed that has been recommended by Someone Else. You decide for yourself what to sell after testing numerous varieties. Trials require that you grow plants far enough apart that each can develop to its full potential. One thing I noticed from doing this was that my trial plots didn't need nearly as much irrigation as my intensive veggie garden. Another was that these well-separated plants got much larger; they tasted better than crowded vegetables did when they weren't harvested promptly; and many vegetable species grown that way yielded more in relation to the space occupied, not less as I had read in books by intensivist gurus.

I sold the seed company in 1986. With lots of free time and several acres of gardenable land to play with, I researched the nearly lost art of vegetable gardening without irrigating at all, which is mainly done by putting plants extremely far apart. Having mastered this, I wrote *Waterwise Vegetables*, a small book about dry gardening in Cascadia (a bioregion encompassing those parts of Oregon and Washington west of the Cascade Mountains, the redwood country of northern California, and the islands and Lower Mainland of British Columbia). It is out of print now.

These days I no longer raise my vegetables using the extreme intensive method that is still advocated by Everybody Else. And I irrigate much less than most people. If I did not have irrigation, I could still grow my garden. I believe I've worked out methods that best suit the coming hard times.

The coming hard times

Several flows are inevitably joining and reinforcing each other, becoming a global river of change:

We are soon going to base our civilization on something other than oil ... or else we aren't going to have much of a civilization left. Soon, everything made with oil is going to cost a lot more: gasoline, food, clothing, transportation, heating of houses, etc. And after that, if oil is still the basis for almost everything we do, then everything is going to cost even more.

At the same time as oil is getting scarcer, average people in the countries with older industrial systems are going to have less real purchasing power. This is the inevitable consequence of a global economic system. Attempts to protect English-speaking labor from competition with Chinese or Indian labor will prove futile and self-destructive. Working people will have less to spend, while much of what they have gotten used to buying in these recent fat years is going to cost more.

Many will respond by producing some or much of their own food. But those practicing raised-bed intensive methods will discover that intensive use of land requires large quantities of water, manure/compost, and fertilizer. If they could make a comparison, they would find that highly intensive beds require more of their time and effort than the slightly increased yield justifies.

Water has become scarce in many places. Flow from rural wells is dropping as many new households pump from the same water table. At the same time, watersheds are becoming ever more degraded, lessening the recharge of groundwater. Electricity will go up in price as oil does, so pumping water will also cost more. Folks on municipal water systems will pay more because both purification and pumping are energy dependent. In short, town water may soon cost so much that even if it is available, and even if watering of gardens is allowed, irrigated gardening as it has been done will induce economic pain.

Fortunately, in most temperate climates, vegetables can be grown with little or no irrigation. Our ancestors knew how to do this in the days before water came out of pipes under pressure. In this book I will show you how to do it as well.

Chemical fertilizers, and many organic ones too, are made with petroleum or natural gas, so they are going to become more expensive. If you can only obtain small amounts of ordinary manure or homebrewed compost in their place, an extensive gardening system will enormously outproduce highly demanding intensive beds. In this book I will show you how to make effective compost simply and without undue effort, and how to wisely choose and use the best manures, avoiding overuse, which is a common gardening mistake. Building up soil excessively not only wastes money and effort, but also lowers the nutritional content of your vegetables.

It's true that there are a few kinds of vegetables, like celery and cauliflower, that require extremely high levels of soil fertility. Rather than try to make super-fertile soil for demanding crops by using manure or compost, I will show you how to concoct an inexpensive organic fertilizer blend made entirely of agricultural waste products and crushed rocks. If this complete organic fertilizer is used to supplement modest amounts of manure and/or compost, the food your garden produces will contain far more (human) nutrition than veggies grown by any other method.

This book is for people who must have a good result. But veggie gardens started with garden-center seedlings and picture-packet seeds often don't have a great result. Seedling raisers and mass-market picture-packet seed businesses are not always ethical. The successful home gardener must start with strong seeds and truly healthy transplants of varieties that are dependable and productive. I was a seedsman. I know the trade. I will tell you who to deal with and why so you'll be a smart buyer and end up with a successful garden. You'll also spend a lot less.

Gardening magazines, garden centers, and seed catalogs all promote the idea that their appealing merchandise is useful and essential — that you need it. Actually, to veggie garden successfully you only need a few hand tools, used properly. I am going to educate you about this as your grandfather should have done. But almost none of us had a grandfather who knew how to grow vegetables, who grew up on a farm, who sharpened shovels and hoes and worked the earth. If you'll allow it, I am going to be the gardening grandfather you never had.

Getting land

To produce a lot of food you have to use a fair bit of land. Not acres, not even a generous half acre, unless you aim for complete family food self-sufficiency, but still a fair bit more than a few postage-stamp beds in a tiny backyard.

People in the United Kingdom experienced hard times from about 1930 to the early 1950s. During World War II, with survival as a nation at stake, they had to do everything possible to produce as much food as could be raised. Every bit of farmland that could be cropped grew essentials like cereals and potatoes; consequently most kinds of vegetables were scarce. So during wartime the country expanded the "allotment," a British term meaning a community garden plot, into a national institution. Every council (local government) was required to make an allotment available to any resident who requested one. By law, each plot, provided for a token rental, had to be at least 300 square yards or 2,700 square feet (about 250 square meters). Some people would take two plots; one for vegetables and the other for small fruit. People would spend time on their allotment after work. Sunday afternoons became gratifying social events at the gardens. Many of the plots sprouted tiny lockable shacks made of recycled materials in which gardeners kept a few tools and an old chair, too, which they could pull out into the sun and rest in as they chatted with the neighbors and sipped a bottle of warm English beer.

Only a few allotments still exist in the United Kingdom because the economy has been good there since the early 80s. But during hard times, having 2,700 square feet of veggie garden made all the difference between health and sickness, between having enough to eat and low-grade hunger.

Another example comes to mind. The Cubans had especially hard times after the fall of the Soviet Union. Before 1991 they had barely been getting by through raising sugar, exporting it to the communist bloc nations, and importing food and gas. The Cubans responded to the collapse of the Soviet Union by breaking up their huge co-op sugar farms and converting them to individual holdings. City people were freely granted garden blocks at the edges of towns and cities; each block was a third of an acre. Today, half of the produce consumed in Havana is grown in urban gardens. And urban gardens produce 60 percent of the vegetables consumed in all of Cuba. Today in Cuba only animal-based foods are scarce. Neighborhood gardens and community horticultural groups not only produce food for their members, but they also donate produce to schools, clinics, and senior centers and still have enough excess produce to sell a bit in neighborhood markets at low prices. By the beginning of the year 2000, there were over 500 community vegetable stands functioning in Cuba, with prices at 30 to 50 percent of the prices at farmers' markets. No one is hungry; the people are well nourished despite ongoing economic sanctions from the US government. What is most interesting to me is that all this produce is organically grown. The Cubans are now on the planet's leading edge in developing holistic non-petroleum-based horticulture.

The community garden has not yet caught on in North America. The ones I supported in Eugene, Oregon, during the 80s and early 90s provided plots that were too small for a serious gardener like me to want to bother working. And I was shocked at that time to see how many unclaimed plots there were. A serious gardener could have rented three or four or ten of them and had a half-decent garden, but no one did that. I guess that's how people think in fat years.

Size of your garden

I wish I could tell you how much food could be produced from a particular amount of land, but there are too many factors at play. How fertile is the land? How deep and what type is the soil? How skilled is the gardener? How much water is available, either from rainfall or from irrigation? How will the weather be that season? What sorts of vegetables will be attempted? Will the crops be ideally suited to the climate and soil? What quality of seeds will be used? What is the latitude (which determines the duration and strength of sunshine for the growing season)? How many frost-free days are there? How severe is the winter?

As a rough gauge, take the 2,700-square-foot wartime allotment plot in the United Kingdom. Britain's cool and frequently cloudy summers mean that most vegetables grow more slowly than they usually do in the United States or southern Canada. But on the plus side, the mild English winters allow gardeners in many areas to harvest frost-hardy crops year-round. The wartime British were not expected to make a complete family diet out of 2,700 square feet of vegetables. Their staff of life was bread from the local baker. They ate as much meat, cheese, and fish as they could get. The children still drank milk from the dairy. Probably during the war years vegetables, including potatoes, did not make up more than a third of the family's total caloric intake.

Tasmania, where I live now, enjoys a slightly milder winter climate than is found in the U.K. This permits me to actively grow root crops, coles (i.e., broccoli, cabbage, cauliflower), and salad greens during all the chilly, frosty months when I can't grow beans, tomatoes, and corn. Over half my garden area produces two crops each year. I use about 2,000 square feet (200 square meters) of actual growing area (not counting paths and surrounds) to supply my kitchen, which feeds a family of two adults (no kids). My garden vegetables make up about half the daily calories we eat year-round. Add in the area used by paths and surrounds and I'm up to nearly the 2,700 square feet of the British allotment.

Presently I can afford (and do conveniently find offered for sale) all the high-quality concentrated organic plant nutrients (fertilizers) my garden can use. If I had to scavenge fertilizer and make compost from all available wastes (possibly including our own humanure), the production of my garden would drop. Drop how much? Maybe by a third if I made good compost; drop by half or more if I made poor compost. (I'll explain making effective compost in Chapter 7.) I could make up for that drop in productivity by using more land — if I had it.

If the earth freezes solid in winter (meaning no winter garden unless it's under plastic or glass), increasing the growing area by half and preserving or (even better) storing produce in a root cellar for winter would do the trick. If irrigation is in short supply or nonexistent in a climate that usually has decent summer rainfall, then increase by half again, to 4,500 square feet (420 square meters), the area needed to produce half the year's calories for two adults. Is it any wonder that the typical small-town building block has, until recently, been a half acre (which is about 21,000 square feet/1,950 square meters)? A lot that size has room for a significant garden in the backyard.

One more thing. If your goal is to produce not half, but nearly all the calories and nutrition needed year-round, and if your family can depend on the ordinary potato as their healthful staff of life, then you can add more land in order to produce sacks and sacks of nutritious spuds or sweet potatoes. If your efforts are helped by irrigation, add at least 500 more square feet (45 square meters) of growing area for each adult in the family. If there is no irrigation and you live in the rainier parts of North America east of the 98th meridian (a north-south line running through Dallas, Texas), add about 750 square feet (70 square meters) per adult. The good thing about potatoes is that working plots of this scale can be done entirely with hand tools. To produce the same amount of nutrition by growing cereal grains would require five to ten times as much land per person. The healthful potato is really the thing for getting through hard times. (If you don't believe that the potato is a health-producing food, please skip forward to Chapter 9 and read what else I have to say about the common spud.)

The healthful spud

The "Irish" potato actually is a native (South) American crop from what is now Peru and Bolivia. However, to arrive in North America it first had to be carried to Spain in the 1500s. For two centuries the Europeans considered it only an amusing oddity until, in about 1700, a plant grown in a botanical garden in Sicily began forming tubers early enough in the season to be useful in temperate latitudes. Then the potato's amazing productivity caused a massive increase of population throughout northern Europe. These people became the migrants who so rapidly filled North American farms and factories.

Before the English brought potatoes to North America, the native North Americans practiced an agricultural system based on corn, beans, squash, and sunflowers, crops that had probably moved into North America from Mexico about 1,000 years before the arrival of the English. This gardening system required nearly an acre to support an extended family. The English-speaking Americans adopted the Native American crops; combined them with food crops brought from England, which included the potato from South America; brought the sweet potato (probably from the Caribbean); and created a hybrid agricultural system combining the benefits of three continents. ■

Is that enough space?

This still may not be enough space. There's one more thing I haven't yet told you. If the soil where you live does not freeze solid in winter to a depth of at least 18 inches for at least a few continuous months, it is probably not possible to grow a vegetable garden on the same land for more than three to five years before serious troubles arise with diseases and/or soil-dwelling insects. Many people living in mild climates have grown the family garden in the same place for more than a generation and think everything is fine. But these folks have forgotten, or were never told by their predecessors, that some kinds of vegetables that once were easy to produce on that plot now seem impossible to grow. They have also forgotten, or never knew, that the output per area used is considerably lower now than it was during the first few years of gardening there.

Winter's freezing halts the soil's biological process. When the thaw comes, the soil ecology starts up again, but from near zero. From this cold start, useful soil microorganisms and small soil animals have as good a chance to dominate as do the unwanted ones. The good guys can be helped out with crop rotation and a bit of compost.

I stated a few paragraphs ago that presently my own garden for two is 2,000 square feet (200 square meters) of irrigated, well-fertilized growing beds. Now I will confess in full: I have nearly two such vegetable gardens. A large section of my land is always resting — not being fertilized, not being watered, growing rough grass and clover like a pasture. The grass areas are roughly mowed a few times each summer, and all the cuttings are allowed to lie in place to decompose. The British name for this practice is a ley. Lasting three to five years, a ley rebuilds the soil's content of organic matter and restores the biological process to a stable, healthy balance. Every four or five years, most of my vegetable beds are put to rest in grass, and the grass beds are turned over and begin to grow vegetables. For the first two years after breaking the sod, veggies on the new ground grow noticeably better than the ones on the old beds were doing. By about the fourth year, the appearance of disease and slower overall growth tells me it is time to rotate again.

Thus the size of my garden has doubled. In my case, from 2,000 square feet of growing beds to over 4,000 square feet (370 square meters). Add paths, a couple of currant bushes, four small fruit trees, some asparagus and perennial herbs around the fringes, and I am using 733 square yards (about 6,500 square feet or 600 square meters) inside a wildlife-proof fence. Why 6,500? Because where I live the fencing comes in rolls of 100 meters (109 yards) and when you wrap 100 meters of fencing around the outside of a square, that is the area you end up enclosing.

Becoming a vegetableatarian

You have to *be* a vegetarian or a vegan. It's an absolute thing. There's no halfway about it. Either you eat meat or you don't. Either you eat dairy and eggs or you don't.

Vegetableatarianism is not like that. First of all, the word itself will not be found in any dictionary because I made it up. When I'd lecture, someone would always ask me if I was a vegetarian. "No," I'd quip, "I'm a vegetableatarian. That's a person who mostly eats mostly vegetables most of the time."

Since vegetableatarianism is my own word, it is a lot like I am — imperfect. I have known for a lot of years how I should eat. I aspire to eat that way.

If I eat in the manner to which I aspire, I will feel better, be healthier, degenerate less (or more slowly), and maybe live longer. But I have bad habits to overcome, pleasurable and self-destructive habits whose allure I have failed to resist an enormous number of times, although slowly, slowly, I show more character.

When I started gardening I was a young puppy of only 31 years old. I had a strong intention then — one that hasn't changed since. My aim was to have food I produced myself make up the largest possible part of my diet, although I never expected to eat only vegetables.

I started down this road as a typical meat-and-potatoes eater. On our first suburban homestead I had rabbits in cages and chickens in a fenced yard under the fruit trees. I even fattened a friendly young steer named Moocow and put his manure into the compost until Moocow's body, perfectly cut and wrapped, went into the freezer.

I didn't stop eating rabbit because killing them was distasteful. It was not regret over murdering poor old Moocow that put me off beef. When I first started gardening, I'd sit down at the table and eat the meat and potatoes first, then fill in any gaps with a bit of broccoli or salad. A few years later I would fill my belly with broccoli and salads and sliced tomatoes ... and then find I was too full to bother with the meat. I had unintentionally become a vegetableatarian. So I sold the sturdy cages I'd built to house the rabbits and was relieved of committing murder wholesale. We did keep a few chickens, but mainly for eggs.

Gardening through my own hard times

In 1978 I was 36 years old, married with no kids, and had a thriving small business, big backyard veggie garden, lots of free time, plenty of money. We decided to drop out of the rat race and homestead, sold almost everything, and bought five acres in a pretty Oregon Coast Range valley. For the first year we adjusted to a new lifestyle — and lived on savings. The next year I wrote and published a gardening book and started a mail-order veggie seed business, both on a shoestring. But the book turned out to be a break-even venture, and Territorial Seed Company didn't start making good profits until mid-1983. By then the savings were long gone.

Between the last half of 1980 and the first half of 1983, what kept me going was the seed company's trials ground. As mentioned earlier, a seed company will test numerous varieties of, in my case, vegetables, before deciding what seeds to sell, and the trials grounds are where these tests take place. My first trials ground was a half-acre plot that grew a lot more vegetables than we could eat. I also continued the family garden to produce kinds of vegetables that would not be tested in the trials that year.

In 1980 and 1981 I mainly ate the trials. In 1982 the business did better. After the sales season ended in June, I decided the business could spare $7,000 for my year's salary. But I had lived through 1981 spending only $4,000 on absolute necessities (and property taxes), so I put $3,000 of that $7,000 into my savings account as a reserve against really hard times and did another year on $4,000. When my once-a-year payday came around again in June 1983, I looked at the books and realized that I had finally created a profitable business. So I began taking $1,000 a month and doubled that the next year. My hard times were over.

From mid-1980 through mid-1983, most of the food my household ate was vegetables, supplemented by some apples from our old orchard and helped out at breakfast most mornings by blackberries, picked during high summer, stored in a rust-speckled old chest freezer in the woodshed, and blended with frozen bananas, bought as "overripes" at super-bargain prices. Money was so tight that when the germination percentage of the seed company's bean seeds dropped below what was ethical to sell, I'd bring those seeds up to the house and we'd cook them. The food we purchased during those years was the odd bit of brown rice or millet, sometimes a chunk of ordinary cheese, some real Jersey butter or milk from the man down the road, olive oil and vinegar for salad dressings, and in winter, oranges or grapefruits now and then, but only by the full box and only when really cheap. I bought enough gasoline to go to town twice a month, paid the land taxes, purchased the odd bit of clothing at the Salvation Army, bought a chunk of beef about once a month when I'd crave it.

The point of this story is that you too can eat that frugally ... if you need to. I could do it again too, if I needed to. And in terms of health we'd both be better off if we did.

From that time up to the day I write these words I am still, by choice, a vegetableatarian. But in these prosperous, easy-living days of the early 21st century, I do not make 80 percent of my diet come from the garden. I can afford to, and do, buy some food I can't grow. To supplement my own apples we'll buy an occasional mango or ripe pineapple from Queensland and the odd avocado at sale price. I can afford the best olive oil and occasionally some mighty fancy cheese. But for health and for the simple joys of eating delicious food, I still grow about half our kitchen's input based on total calories consumed, and more than half measured by cost. You can do that or more, and you will be far better for it if you do.

CHAPTER 2
Basics

I assume many of you are reading my book because you seriously need to make a food garden, starting just as soon as you can put some seeds or seedlings into the earth. I assume you can't afford costly mistakes and wasted efforts. But these days, few people have had the good fortune to grow up on the land, so new vegetable gardeners have to catch up on some basics.

What is a vegetable?

Every vegetable we eat was once a barely edible wild plant. As we humans improved that wild plant, its ability to compete was reduced. Vegetables have become oversized weaklings unable to best their ancestors, who remain the wiry guerilla fighters nature demands of its survivors.

A landscape may appear beautiful and serene, but look closely: a fierce battle is going on. Each plant is trying to overcome or outwit its neighbor, struggling to control light, moisture, and space. The scene seems peaceful only because the action is in slow motion and you are seeing the winners of the moment.

Our vegetables lack the survival skills wild plants still have. Wild plants know how to toughen up and conserve moisture when the soil gets dry; a moisture-stressed vegetable becomes nearly inedible. Wild plants will cope if they are put into the shade; veggie gardens must be grown in full sun or at least have the sun from 10 am until 4 pm. When the earth is infertile, wild plants adapt. Put a vegetable into the same circumstances and it usually just dies. Wild plants protect themselves from animals and insects with intensely bitter, unpleasant flavors. They make themselves poisonous. But we humans wish to eat sweet, pleasant-tasting food. So we must protect our relatively helpless veggies.

Wild plants usually make thousands of tiny, hard seeds so that even if most of these seeds are eaten or, having fallen on hard times, die, at least a few will manage to start the next generation. But we must sow vegetable seeds in welcoming conditions where they will survive and grow.

Wild plants have leaves and stems filled with strong fibers that resist damage from wind and heavy rain. Their roots anchor the plant firmly. But we humans want to eat tender juicy leaves, stems, and roots. If a veggie garden bears the brunt of a gale, leaves will likely be shredded and many of the plants will be half-uprooted. Veggie gardens require sheltered spots.

Most wild plants grow a huge network of roots to aggressively mine the soil. These roots are formed at the expense of making leaves. This is a pro-survival trade-off. But humans usually want the tops to grow large and lush, and this can only be done at the expense of weakening the root system, a reverse trade-off.

Luther Burbank, the famous 19th-century plant breeder, suggested that when humans domesticated a wild species, it was as though we made an enduring contract reading something like this: You agree to be my plant and to let me change you into something I will like better than the way you are now. I, in turn, promise that instead of leaving you to struggle to survive, I will eliminate or greatly reduce the competition you face. I will put you into much better circumstances than you now find in nature — moister, looser, richer soil — and I will value your seeds and propagate them forever. With this contract in hand, we began to alter our plants, changing them into what we wanted. We encouraged them to make bigger, juicer leaves (lettuce, spinach); larger flowers (broccoli, cauliflower, globe artichokes); or fewer but fat, tender, more flavorsome seeds (beans, wheat). Domesticated plants have greatly enlarged the upper storage chamber of their root while reducing the extent of the deeper subsoil system (carrots, beets, parsnips, potatoes). We caused them to make much larger, better-tasting (and fewer) fruit with thinner skins and fewer seeds (tomato, chili, eggplant). Burbank also said that whatever we envisioned plants becoming, they did, changing themselves to please us because we humans had become their protectors.

Although it took thousands of years to breed the changes we made in our food crops, it wasn't too hard to accomplish because domesticatable plants are like dogs wagging their tails, trying to please. Our main breeding technique has been to first bare and loosen the soil, then place our seeds far enough apart that the plants don't compete too much with each other. In this circumstance our crop benefits from all the water, light, and soil nutrients it can use, and each plant has enough room to develop well. If we happen to put in too much seed and our crop is too crowded, competing with itself, we remove some of these plants. This is called *thinning*. We also eliminate any wild plants trying to sneak in. This is called *weeding*. And then we save seeds from those plants that most please us. This is plant breeding in a nutshell.

These few simple practices are almost all there is to agriculture or to vegetable gardening: We eliminate wild plants; put in the ones we want to grow; space our plants so that they aren't overly competing with each other; keep the wild plants from taking over; make our garden soil more fertile and more moist than nature does. And then our gentled plants bless us by happily allowing us to eat them because they are confident we are going to carry on their progeny in the coming years.

Field crop or vegetable

"Field crops" include the cereal grains — wheat, rye, oats, and barley — as well as corn (maize), rice, sorghum, millet, quinoa, and amaranth; some varieties of beans and peas, usually the types grown for dry seed; and oilseeds such as flax, sunflower, sesame, poppy, and canola. A field crop can be productive in unfertilized, ordinary soil, but it does better when that soil has been improved. Heirloom varieties, bred before the use of chemical fertilizers, are especially good at growing in ordinary soil.

Vegetables are different. For thousands of years the kitchen garden has received the best of the family's manures and lots of them. After millennia of coddling in highly improved conditions, few veggies thrive in ordinary soil.

Low-demand vegetables. Some vegetable species can still cope with soil of the sort that will grow field crops if the soil has been well loosened by spading or rototilling. Low-demand vegetables include carrots, parsnips, beets, endive/escarole, snap and climbing (French) beans, fava beans (broad beans), and garden peas (see the sidebar listing vegetables by the level of attention they require). However, when low-demand vegetables are given soil considerably more fertile than their minimum requirement, they become far more productive. Low-demand crops are capable of struggling along and usually will produce something edible even under poor conditions.

Vegetables by level of care needed

Low Demand	Medium Demand	High Demand
Jerusalem artichoke	globe artichoke	asparagus
beans, peas	basil, cilantro (coriander)	Italian broccoli
beet	sprouting broccoli	Brussels sprouts (early)
burdock	Brussels sprouts (late)	Chinese cabbage
carrot	cabbage (large, late)	cabbage (small, early)
other chicories	cutting (seasoning) celery	cantaloupe/honeydew
collard greens	sweet corn	cauliflower
endive	ordinary cucumbers	celery/celeriac
escarole	eggplant (aubergine)	Asian cucumbers
fava beans	garlic	kohlrabi (spring)
herbs, most kinds	giant kohlrabi	leeks
kale	kohlrabi (autumn)	mustard greens (spring)
parsnip	lettuce	onions (bulbing)
southern peas	mustard greens (autumn)	peppers (large fruited)
rabb (rappini)	okra	spinach (spring)
rocket (arugula)	potato onions	turnips (spring)
salsify	topsetting onions	
scorzonera	parsley/root parsley	
French sorrel	peppers (small fruited)	
Swiss chard (silverbeet)	potatoes (sweet or "Irish")	
turnip greens	radish, salad and winter	
	rutabaga (swede)	
	scallions (spring onions)	
	spinach (autumn)	
	squash (pumpkin), zucchini	
	tomatoes	
	turnips (autumn)	
	watermelon	

■

Medium-demand vegetables. These veggies need significantly enriched soil to thrive. This group includes lettuce, potatoes, tomatoes, corn, etc. What I consider the minimum enrichment for this group would be spading in a bit of agricultural lime plus either a half-inch-thick (12-millimeter) layer of well-rotted low-potency manure or else a quarter-inch-thick (6-millimeter) layer of well-made potent compost. But medium-demand vegetables will do enormously better when given soil considerably more fertile than their minimum requirement.

High-demand vegetables. These are sensitive, delicate species. High-demand vegetables usually will not thrive unless grown in light, loose, always moist soil that provides the highest level of nutrition. These vegetables become rather inedible unless they grow rapidly. In Chapter 9 I will discuss these vegetables one by one and will provide full details about how to cultivate them. If gardeners are not able to provide the nearly ideal conditions required by high-demand vegetables, they would be much better off not attempting to grow them.

Helping plants grow

Plants need minerals to construct their bodies. So do people. Human bodies need energy and enzymes, minerals, vitamins, amino acids, and, as research into nutrition continues, it seems an ever-extending list of other substances. Plants also need a wide assortment of organic chemicals produced by the soil ecology. These were named "phytamins" by the Russian researcher Krasil'nikov, who was, in my opinion, the world's greatest soil microbiologist. The creation of phytamins is accomplished by microorganisms, and the process is still not well understood.

However, the mineral nutrition of plants is more straightforward. Agronomists are confident about which minerals are required, and in what proportions. As an example, most plants use a lot of calcium, but for every six to eight measures of calcium, they'll also need one measure of magnesium, maybe a sixteenth measure of sulfur, and one ten-thousandth measure of boron. If they have heaps of calcium but are short of magnesium, then they won't be able to grow any more than the amount allowed by the quantity of magnesium they've got. If they have adequate calcium, magnesium, and sulfur, all in the right proportions for ideal growth, but are desperately short of

Figure 2.1: *Each stave in the barrel represents a different plant nutrient in soil. The ones at their full height are in full supply. The barrel at the left illustrates a soil in which the greatest deficiency is of nitrogen; plants will grow no better than the level of nitrogen. The barrel on the right shows that soil after nitrogen has been added. Now the most limiting deficient nutrient is potassium. The plants will grow no better than the level of potassium permits.*

boron, then they will grow as poorly as though they were short of calcium and magnesium and sulfur.

According to Dr. William Albrecht, chairman of the soils department of the University of Missouri during the 1940s and 1950s, the broadest differences in soils are caused by climate. Albrecht repeatedly pointed out that where it rains just barely enough for agriculture, the soils tend to be the richest and most balanced. Usually prairie grasses grow there. Where it rains a great deal, where vegetation looks lush and trees grow thickly, the soil is usually, overall, much less fertile and much less balanced. Sparsely grassed prairies once supported huge herds of healthy animals. Food from these fertile, balanced soils makes people healthy too. Before the trees were cleared, forested lands supported relatively few animals that tended to be smaller, less healthy, with a shorter lifespan than those on the prairies.

The best I can say about these well-rained-on soils is that they aren't too bad for growing something. What do I mean by "not too bad"?

Before World War II most North Americans got most of their food from farms located not far from where they lived. As a result, differences in average human health due to local soil conditions were apparent. Albrecht provided this example: In 1940 the United States instituted a draft registration for military service. All young men had to report for a physical examination. In Missouri, the prairie soils in the northwest are far more fertile than the once thickly forested soils in the southeast of that state. If you draw a line across a map of Missouri from northwest to southeast, and test the soils all along that line, you'll find that they get progressively poorer as you travel toward the Mississippi river, precisely as the amount of annual rainfall increases along that same line. Accordingly, approximately 200 men out of 1,000 examined from the northwest of Missouri were found to be unfit for military service, while 400 young men out of 1,000 from the southeast of the state were unfit. In the center of Missouri, about 300 per 1,000 young men were unfit.

Suppose the soil in your area contains abundant mineral nutrients in a near-perfect balance. In that case, the plants grown in your district will be healthy. If you take the manure from the animals eating that vegetation and/or take the vegetation itself and rot it down into compost, and then spread that compost or that manure atop your garden, what you have done is transported minerals *in the right proportions* to your garden and increased the overall level of minerals in the garden's soil *in the right proportions.* The result is a much better garden producing the same highly nutritious food the surrounding land grows.

But suppose that the soils in your area do not contain a perfect balance of plant nutrients. This means the average vegetation in your area is not as healthy as it might be, and neither are the animals grazing on it. When you bring a load of that vegetation into your garden in the form of manure or compost, you are increasing the amount of organic matter in your soil, which is good. You are also increasing the amount of minerals in the soil, but you are amplifying the imbalances of those minerals. Importing large quantities of imbalanced organic matter often leads to trouble because rich and ideally balanced soils are rare, not common. When you depend on your garden to provide much of your food, your health will suffer to the degree that your soil is out of balance. Homesteaders mainly eating out of their gardens, even from organically grown gardens, have developed severe dental conditions or other health problems as a consequence of incorrect soil building.

If the soils around you are not rich and balanced, you should, if possible, take steps beyond ordinary manuring or composting to improve them.

Complete organic fertilizer

Because my garden supplies about half of my family's yearly food intake, I maximize my vegetables' nutritional quality. Based on considerable research, I formulated an organic soil amendment that is correct for almost any food garden. It is a complete, highly potent, and correctly balanced fertilizing mix made entirely of natural substances, a complete organic fertilizer, or COF. I use only COF and regular small additions of compost. Together they produce incredible results. I recommend this system to you as I've been recommending it in my gardening books for 20 years. No one has ever written back to me about COF saying anything but "Thank you, Steve. My garden has never grown so well; the plants have never been so large and healthy; the food never tasted so good."

COF is always inexpensive judged by the results it produces, but it is only inexpensive in money terms if you buy the ingredients in bulk sacks from the right sources. Finding a proper supplier will take urban gardeners a bit of research. Farm and ranch stores and feed and grain dealers are the proper sources because most of the materials going into COF are used to feed livestock. If you should find COF ingredients at a typical garden shop, they will almost inevitably be offered in small quantities at prohibitively high prices per pound. If I were an urban gardener, I would visit the country once every year or two and stock up.

A fertilizer that puts the highest nutritional content into vegetables provides plant nutrients in the following proportions: about 5 percent nitrate-nitrogen (N), 5 percent phosphorus (P), and only 1 percent potassium (K, from kalium, its Latin name). It would also supply substantial and perfectly balanced amounts of calcium (Ca) and magnesium (Mg), and minute quantities of all the other essential mineral plant nutrients such as iodine, cobalt, manganese, boron, etc. The ideal fertilizer would release slowly, so the nutrients didn't wash out of the topsoil with the first excessive irrigation or heavy rain. It would be a dry, odorless, finely powdered, completely organic material that would not burn leaves if sprinkled on them and would not poison plants or soil life if somewhat overapplied. All this accurately describes COF.

You could sizably increase bulk yield by boosting the amount of potassium in garden soil higher than COF will, but the nutritional content of the veggies would decrease by just about as much as the yield went up. Most commercial growers, be they chemical or organic growers, push soil potassium to high levels for the sake of profit. But the higher bulk yield potassium triggers is in the form of calories — starch and fiber — and not in the form of protein, vitamins, enzymes, and minerals, which we need a lot more from our food.

COF is concocted by the gardener. All materials are measured out by volume: that is by the scoop, bucketful, jar full, etc. Proportions varying plus or minus 10 percent of the targeted volume will work out to be exact enough. *Do not attempt to make this formula by weight.*

I blend mine in a 20-quart (20-liter) white plastic bucket using an old 1-quart (1-liter) saucepan for a measuring scoop. I make 7 to 14 quarts (7 to 14 liters) of COF at a time. The formula is shown in Figure 2.2.

Formula for complete organic fertilizer

Mix fairly uniformly in parts by volume:

4 parts any kind of seedmeal except coprameal

OR

3 parts any seedmeal except coprameal and 1 part "tankage" (sometimes called "blood-and-bone" or "meatmeal"). This higher-nitrogen option is slightly better for leafy crops in spring

OR

4½ parts less-potent coprameal, supplemented with 1½ parts tankage to boost the nitrogen content

BLEND WITH

¼ part ordinary agricultural lime, best finely ground

AND

¼ part gypsum (if you don't use gypsum, double the quantity of agricultural lime)

AND

½ part dolomite lime

PLUS (for the best results)

1 part of any one of these phosphorus sources: finely ground rock phosphate (there are two equally useful kinds or rock phosphate, "hard" or "soft"), bonemeal, or high-phosphate guano

½ to 1 part kelpmeal or 1 part basalt dust

Figure 2.2

Seedmeal and the limes are the most important ingredients. These items alone will grow a great-looking garden. Gypsum is the least necessary kind of lime and is included because it also contains sulfur, a vital plant nutrient that occasionally is deficient in some soils. If gypsum should prove hard to find or seems too costly, don't worry about it and double the quantity of inexpensive agricultural lime. If you can afford only one bag of lime, on most soils, in most circumstances, your best choice would be dolomite, which contains more or less equal parts of calcium and magnesium. You could also alternate agricultural lime and dolomite from year to year or bag to bag.

Guano, rock phosphate, bonemeal, and kelpmeal may seem costly or be hard to obtain, but including them adds considerable fortitude to the plants and greatly increases the nutritional content of your vegetables. Go as far down the list as you can afford to, but if you can't find the more exotic materials toward the bottom of the list, I wouldn't worry too much. However, if shortage of money is a concern that stops you from obtaining kelpmeal, rock dust, or a phosphate supplement, I suggest taking a hard look at your priorities. In my opinion, you can't spend too much money creating maximum nutrition in your food; a dollar spent here saves several in terms of health costs of all sorts — and how do you place a money value on suffering?

Seedmeals. Seedmeals, a by-product of making vegetable oil, are mainly used as animal feed. They are made from soybeans, linseed, sunflowers, cottonseeds, canolaseed (rape), etc. When chemically analyzed, most seedmeals show a similar NPK content — about 6-4-2 — although coprameal (which is made from coconuts after the coconut oil is squeezed out and is used mainly for conditioning race horses) is about one third weaker in terms of NPK than the other seedmeals. Coprameal has a plus point, however: coconuts are almost inevitably grown without pesticides or chemical fertilizers.

The content of minor nutrients — calcium, magnesium, and trace nutrient minerals — varies quite a bit by kind and from purchase to purchase, depending mostly upon the soil quality that produced the oilseeds. Most farm soils are severely depleted, so most seedmeals in commercial trade are probably rather poor in terms of supplying nutrients other than NPK. And because seedmeals are intended as feed and not as fertilizer, they are labeled by their protein content, not by their content of NPK. The general rule is that for each 6 percent protein, you are getting about 1 percent nitrogen, so buy whichever type of seedmeal gives you the largest amount of protein for the least cost. Seedmeals are stable and will store for years if kept dry and protected from mice in a metal garbage can or empty oil drum with a tight lid.

Tankage. Also known as meatmeal, tankage is a product of the slaughterhouse. In the United States it is made up of the whole animal, minus the bones and fat, ground to a powder and dried. In Australia a similar product is called "blood and bone" and contains everything but the fats. Tankage has no odor as long as it is kept dry, and it produces no odor when spread on the earth as part of COF. I have no health concerns regarding meatmeal, and I've used it for more than 20 years. However, if you're worried about disease and decide not to use tankage, the only consequence will be that your COF will be slightly lower in nitrates, requiring that you use a bit more of it to get the same growth response. The form of tankage I get now analyzes at 10-4-0.

Lime(s). There are three types of lime, which is ground, natural rock containing large amounts of calcium. "Agricultural lime" is relatively pure calcium carbonate. "Dolomitic lime" contains both calcium and magnesium carbonates, usually in more or less equal amounts. "Gypsum" is calcium sulfate. (Do not use quicklime, burnt lime, hydrated lime, or other chemically active "hot" limes.) If you have to choose only one kind, it probably should be dolomite, but you'd get a far better result using a mixture of the three types. These substances are not expensive if bought in large sacks from agricultural suppliers.

You may have read that the acidity or pH of soil should be corrected by liming. I suggest that you forget about pH. Liming to adjust soil pH may be useful in large-scale farming, but is not of concern in an organic garden. In fact, the whole concept of soil pH is controversial. (If you are interested in the debate, read the papers of William Albrecht cited in the Bibliography.) My conclusion on the subject is this: if a soil test shows your garden's pH is low and you are advised to lime to correct it — don't. Each year just add what I recommend in the sidebar "Soil improving in a nutshell": compost/manure and 50 pounds of lime(s) per 1,000 square feet (25 kilograms per 100 square meters), or else COF. Over time the pH will correct itself, more because of the added organic matter than from adding calcium and/or magnesium. If your garden's pH tests as acceptable, use my full recommendation in "Soil improving in a nutshell" anyway because vegetables still need calcium and magnesium as nutrients and in the right balance.

One final thing about limes. If you routinely garden with COF, there will no longer be any need to lime the garden. COF is formulated so that when used in the recommended amount, it automatically distributes about 50 pounds of lime(s) per 1,000 square feet per year.

Phosphate rock, bonemeal, or guano. All three of these boost the phosphorus level and (except bonemeal) are usually rich in trace elements. Seedmeals of all sorts once had quite a bit more phosphorus in them, but over the last 25 years their average phosphorus content has steadily decreased. Incidentally, a comparison of nutritional tables published by the USDA over the past 25 years (covered in the March 2001 issue of *Life Extension Magazine*) shows that the average nutritional content of vegetables has also declined about 25 to 33 percent across the board — all vegetables, all vitamins and minerals. This is one reason I grow my own.

Kelpmeal. Kelpmeal has become shockingly expensive, probably in response to environmental degradation, but one 55 pound (25 kilogram) sack mixed into COF will last the 2,000-square-foot (200-square-meter) gardener several years. Kelp supplies plants with some things nothing else will supply — a complete range of trace minerals plus growth regulators and natural hormones that act like plant vitamins, increasing resistance to cold, frost, and other stresses. The kelpmeal imported from Korea costs less than the meal imported from Norway, but as far as I know, kelp is kelp is kelp. So long as the sack isn't half sand, the price difference is only the cost of labor and the exchange rate.

There is a way to economize with kelp. Instead of putting dry meal into the soil, you can spray a tea made with kelp on the leaves. This is called "foliar feeding." There are prepared kelp teas, but you can also make your own. Foliar feeding is more time-consuming than mixing the kelp into COF, but I believe the practice has advantages beyond frugality. Since foliar feeding needs to be done at least every two weeks, the spraying gets the gardener walking among the plants, putting into practice the old saying that the best fertilizer of all is the feet of the gardener.

Some rock dusts are highly mineralized and contain a broad and complete range of minor plant nutrients. These may be substituted for kelpmeal. But I believe kelp is best.

Using complete organic fertilizer

Preplant. Before planting each crop, or at least once a year (best in spring), uniformly broadcast four to six quarts (four to six liters) of COF atop each 100 square feet (10 square meters) of raised bed or down each 50 feet (15 meters) of planting row in a band 12 to 18 inches (30 to 45 centimeters) wide.

Blend the fertilizer in with hoe or spade. If, despite my soon-to-come advice to the contrary, you do not dig your garden, just spread it on top of everything. Soil animals will eat it and mix it in for you. If you're making hills, mix an additional teacup of COF into each. This amount provides a degree of fertility sufficient for what I've classed as "low-demand" vegetables — like carrots, beets, parsley, beans, peas — to grow to their maximum potential. It is usually enough to adequately feed all "medium-demand" vegetables. If you're using less-potent coprameal as the basis of your mixture, err on the side of generous application.

Side-dress medium- and high-demand vegetables. If you want maximum results, then, in addition to the basic soil-fertility-building steps, I suggest that a few weeks after seedlings have come up or been transplanted out, sprinkle small amounts of fertilizer around them, thinly covering the area that the root

Figure 2.3: *Side-dressing plants.*

system will grow into during the next few weeks. As the plants grow, repeatedly side-dress them every three to four weeks, placing each dusting farther from the plants' centers. Each side-dressing will require spreading more fertilizer than the previous one. As a rough guide to how much to use, side-dress about four to six additional quarts (four to six liters) per 100 square feet (10 square meters) of bed, in total, during a full crop cycle. If a side-dressing fails to increase the growth rate over the next few weeks, that lack of result indicates it wasn't needed, so do it no more.

Chemical fertilizers

Concentrated sources of pure plant nutrients should come with a label warning about giving plants too much. That is one reason I don't recommend the use of chemical fertilizers. It is too easy for the inexperienced gardener to cross the line between enough and too much.

Another word of warning: You can poison plants with chicken manure and with seedmeal, too. They are powerful but not nearly so dangerous as chemicals because they are slow to release their nutrient content.

Chemical fertilizers are too pure. Unless the manufacturer's intention was to put in tiny amounts of numerous other essential minerals, the chemical sack won't supply them. What is especially troublesome is that chemical fertilizers rarely have any calcium or magnesium in them. This is particularly true of inexpensive chemical blends, so-called complete chemical fertilizers like 16-16-16 or 20-20-20 or 10-20-16. They are entirely incomplete. They only supply nitrates, phosphorus, and potassium. Depending on how the fertilizer was concocted, there may also be plenty of sulfur (S) in them, but not necessarily. Plants do need NPK (and S), but they also need large amounts of calcium and magnesium, and a long list of other essential minerals in tiny traces. Plants lacking any essential nutrients are more easily attacked by insects and diseases, contain less nourishment for you, and often don't grow as large or as fast as they might.

There is yet another negative to mention. All inexpensive chemical fertilizers dissolve quickly in soil. This usually results in a rapid boost in plant growth, followed five or six weeks later by a big sag, requiring yet another application. Should it rain hard enough for a fair amount of water to pass through the soil, then the chemicals already dissolved in the soil water may be transported as deeply into the earth as the water penetrates (this is called "leaching"), so deep that the plant's roots can't access it. With one heavy rain or one too-heavy watering, your topsoil goes from being fertile to infertile. The risk of leaching is especially great with soils that contain little or no clay.

Organic fertilizers, manures, and composts, on the other hand, release their nutrient content only as they decompose — as they are slowly broken down by the complex ecology of living creatures in the soil. The length of this process is determined by the soil temperature. The rate of decomposition roughly doubles for each 10°F (5°C) increase of soil temperature. Complete decomposition of COF takes around two months in warm soil. During that entire time, nutrients are steadily being released. Gardeners in hot climates (with warmer soils) will get a bigger result from smaller applications of less-potent organic materials. The result will last a shorter time, too, though.

Chemical fertilizers can be made to be "slow release," but these sorts cost several times as much as the type that dissolves rapidly in water. Slow-release chemicals are used to grow highly valuable potted plants and other nursery stock. Seedmeals, naturally slow releasing, are cheaper.

Manures and simple fertilizers

As mentioned earlier, spreading manure and compost made from vegetation in the district you live in is no guarantee of making garden soil that will produce the most highly nutritious vegetables possible. It will make nutritious food only if the soils around you are rich and balanced. Figure 2.4 shows the range of fertilizing values manures and simple fertilizers might have, depending on the nutrient balance in the feed the horse, cow, sheep, etc., was fed.

Please notice that horse manure can range from 0.5 percent nitrogen to 2.5 percent N. Cow manure can go as high as 2 percent, but may be as poor as 0.5 percent. The same spread occurs for phosphorus and potassium. There are two probable reasons for these wide ranges of value. One is that there is no telling how much, if any, bedding material was mixed into the samples being tested. More importantly, it is absolutely certain that what the animals ate varied greatly.

An acquaintance, I'll call him Ken, grows a seemingly excellent garden using horse manure. (I'm somewhat skeptical because his children's teeth and jawbones haven't developed as one would expect for people growing up eating mainly organically grown vegetables). He uses nothing but what I consider enormous overdoses of horse manure and some agricultural lime. Others around me also use horse manure and get poor results. What is going on? Ken gets his manure at no cost from a nearby stable owned by a wealthy man who breeds, raises, and trains thoroughbred racehorses. These animals are fed like

Fertilizing values of manures and simple fertilizers

Kind of manure	N (percent)	P (percent)	K (percent)	C/N (ratio)
Dairy manure	0.5–2.0	0.2–0.9	0.5–1.5	12–25
Feedlot manure	1.0–2.5	0.9–1.6	2.4–3.6	
Horse manure	0.5–2.5	0.3–2.5	0.5–3.0	10–25
Sheep manure	1.0–4.0	1.0–2.5	1.0–3.0	
Rabbit manure	0.8–2.8	1.0–3.7	0.2–1.3	
Poultry manure	1.1–6.0	0.5–4.0	0.5–3.0	6–10
Urine	15–18			0.8
Humanure	5–7	3–5.4	1.0–2.5	5–10
Hog manure	0.3–0.5	0.2–0.4	0.4–0.5	
Duck manure	0.6	1.4	0.9	
Guano, bat	1–8	4–10	0–1	
Guano, seabird	1–12	8–12	1–3	
Fishmeal	10.0–12.0	3.2–6.0	0.0–0.8	4
Seedmeal	4–8	1.2–4.0	1.0–2.0	5–7
Feathermeal	14	0.8	0.3	
Meat and bone	9.0	5.5	1.4	
Dried blood	12.0–14.5	0.4–1.5	0.6–1.0	3

NOTE: These figures were gathered over many years from numerous sources. I would be dubious about any row showing a single figure. If I am told that duck manure has 0.6 percent nitrogen, I don't believe it. If I am told it has from 0.4 to 1.3 percent nitrogen, I believe I'm being told the possible range for duck manure, at least as so far discovered. The extraordinary range in guanos is because some are fresh deposits and some are ancient fossilized rock-like materials that have lost all or virtually all their nitrogen.

Figure 2.4

endurance athletes. Their hay is purchased from carefully selected fields, grown by knowledgeable farmers who know exactly when to cut it for the highest protein content and how to dry it so as to preserve the value. These horses are also given a broad range of protein, vitamin, and mineral supplements, including oilseed meals and abundant kelpmeal. Is it any wonder this manure grows stuff, even if Ken is using too much and throwing his soil out of chemical balance?

Another neighbor sells horse manure from three sorry-looking nags loafing in her back pasture. The vegetation on their compacted, scabby field is pathetic despite all the horse urine the pasture receives. Because there is so little grass coming from that field, she has to buy most of the horses' food. She feeds mainly local grass hay of a quality so low that the racehorse breeder I mentioned wouldn't use it for anything but bedding straw. The manure makes rose bushes, low-demand plants, grow a little better.

My point? It is not easy to obtain manure that you can be confident about — that you can be sure will grow a great garden and produce highly nutritious food. That's why I recommend using only the minimum amount of manure necessary to supply the soil ecology's requirement for organic matter and suggest using COF to supply the plant's requirement for nutrients.

Increasing soil fertility

Some of my readers will be growing a food garden without much money to spend on it. Some may not have access to free or almost free manure (except humanure). Some may not have the means to haul enough animal manure or materials to make an adequate amount of compost. Some will be more affluent; to them the idea of spending five or ten cents to grow a dollar's worth of food will seem entirely reasonable and doable. Because of this range of readers, I will suggest a sequence of gradient steps instead of describing only one approach to increasing soil fertility. By gradient, I mean that each step is a step up and will give you a better result.

No soil improvers. If all I could manage to do with a piece of ground was spade it up, keep the weeds hoed out, and hope for the best, then all I would attempt there would be low-demand crops. Frankly, I can't envision people in this situation unless they are starting a new garden without the first completed compost heap and cannot face the thought of using humanure.

Insufficient soil improvers available. Soil improvement materials were always scarce for Native American gardeners. Their approach, and a wise one it was: concentrate what fertilizer they could get into hills. Usually several plants were grown on each fertilized hill. The word "hill" in this case does not mean a high mound; it is a low broad bump about 18 inches (45 centimeters) in diameter that is deeply dug up and into which fertilizer is mixed. For Native Americans the fertilizer was often a buried bony "trash" fish that weighed at least one pound (500 grams). I expect the Native American garden was also used as the family latrine and as the burial yard for slaughtering waste, dead dogs, and other small animals, with each deposit making a hill.

These days, hills might be fertilized by roadkill or a quart (liter) of strong composted chicken manure and a few tablespoonfuls of agricultural lime. Used that frugally, one bag of chicken manure might fertilize quite a few hills, thus producing an enormous amount of food compared to the cost of the manure. If I had to fertilize hills with raw manure, I'd give the hill at least a month to mellow before sowing seeds. If I were collecting raw manure, I'd also try to collect urine and pour a pint or two into each hill.

A garden grown this way usually concentrates on raising large-sized low- or medium-demand vegetables such as squash or tomato vines. Such a hill could also grow a couple of corn plants, a cucumber vine, or a few kale or collard plants. The hills are always separated by about four feet (just over a meter) from center to center. I am guessing about this next statement of mine — it is based on only one reference in a document — but I suspect that the Native Americans east of the Mississippi, lacking steel shovels, did not dig up the ground between the hills. Instead they kept it under a permanent leaf mulch, bringing in loads of forest duff and spreading it between the hills.

Minimally sufficient soil improvers available. The next step up in both effort and cost would be to make highly fertile hills as before, but also to dig up the soil between the hills and fertilize it less intensely than you do the hills. This means the plants on those hills could more effectively extend root systems and find more nutrition beyond the hills they start out in. Gardening this way naturally leads into forming wide beds or wide rows for small plants while making extra fertile spots in these beds or rows for individual large plants.

In climates where there is good rainfall during the growing season, such as east of the 98th meridian in the United States, gardeners should spread agricultural lime (or for an even better nutritional outcome, a mixture of half agricultural lime and half dolomite lime) at a (combined) rate of 50 pounds per 1,000 square feet (25 kilograms per 100 square meters) before spading the garden. You don't need a soil test for this step; light liming is always a safe and productive thing to do as long as your climate is humid enough to grow a lush forest. If the climate is semi-arid, the soil likely won't need more calcium. Adding needed lime will greatly increase the health of your plants and boost the overall output and nutritional quality of all the food you are growing. Agricultural lime will especially make everything in the cabbage family grow better. If your soil is a bit short of magnesium, dolomite lime contains enough calcium to help the cabbages, while the magnesium it contains will make most of your vegetables taste a lot sweeter.

If your soil already has sufficient calcium and magnesium, the small amount I am suggesting you add won't throw things out of balance. Finding out by soil test if a veggie garden needs more calcium and magnesium is usually a waste of money. An accurate and meaningfully interpreted soil test will cost far more than a little lime. However, for a farmer with a large area to manage, a proper soil test is a relatively small outlay compared to the cost of putting unnecessary lime (or other things) into many acres.

Spread lime once a year, either in spring or in autumn. Uniformly distributing 50 pounds of lime(s) per 1,000 square feet is quite difficult because it makes an incredibly spotty dusting. Spreading more, say 100 pounds per 1,000 square feet, which amounts to two tons per acre, is a lot easier, but even when the application goes up above three tons per acre, you barely turn the soil white with lime dust. However, spreading more than one ton per acre can be dangerous; over-liming ruins plant growth for years afterward. Regrettably, I've done it; don't you do it, too. As my father said, the cheapest experience you can get comes secondhand — if you'll buy it.

There are a few soils (usually clays in arid places) that need gypsum and do not receive much benefit from agricultural lime. There are also a few soils that need to avoid additional magnesium entirely because they already have too much of it, verging on magnesium toxicity (your local agricultural extension service will know if there are such soils in the area). But if you live where dense forests grow naturally and where it rains more than 30 inches (75 centimeters) per year, you probably have nothing about liming to be concerned about except making sure you don't overdo it.

Another equally useful step is to minimally cover the entire garden with manure or compost at least a quarter inch (six millimeters) deep, and preferably double that depth, before spading it up. If I did not have enough well-decomposed ruminant manure or compost to cover the entire garden, I would manure whatever part of the area I could and use that area for growing both low- and medium-demand veggies. I could also attempt low-demand vegetables on those areas that I limed but did not manure; if the soil is naturally fairly fertile, the unmanured soil might still grow something useful. A somewhat better approach in an ongoing garden would be to grow low-demand vegetables on the ground that had been given manure or compost the previous year because some of that added fertility will still be in the soil the next growing season.

Soil improving in a nutshell

You will almost certainly have a reasonably productive garden if once a year (usually in spring), before planting crops, you spread and dig in the following materials for the type of vegetables you are growing.

Low-Demand Veggies	Medium-Demand Veggies	High-Demand Veggies
¼″ (6 mm) layer sacked steer manure or ¼″ finished compost	¼″ (6 mm) layer sacked steer manure or ¼″ finished compost	½″ (12 mm) layer sacked steer manure or ½″ finished compost
50 lb/1,000 square ft (25 kg/100 m²) lime(s) spade, rake, plant	⅛″ (3 mm) layer composted chicken manure	¼″ (6 mm) layer composted chicken manure
	50 lb/1,000 square ft (25 kg/100 m²) lime(s) spade, rake, plant	50 lb/1,000 square ft (25 kg/100 m²) lime(s) spade, rake, plant
	OR (this is far better)	
¼″ layer steer manure or ¼″ finished compost	¼″ layer steer manure or ¼″ finished compost	½″ layer steer manure or ½″ finished compost
4 qt COF/100 square ft (4 L/10 m²) spade, rake, plant	4 to 6 qt. COF/100 square ft (4–6 L/10 m²) spade, rake, plant	4 to 6 qt. COF/100 square ft (4–6 L/10 m²) spade, rake, plant ☞

For large plants I would make hills, as described in the previous section, giving each hill additional strong fertilizer. Agronomists call this practice "banding," the idea being that as the young plants put out roots, they immediately discover a zone or band of highly fertile soil. This help gets them growing fast from the beginning so they outgrow environmental menaces. (One insect can take what seems a huge chunk out of a newly emerged seedling and may kill it from a single feeding. But once that plant becomes 50 times larger, the removal of the same single chunk seems but a tiny insult.) Later on, having had a good start, they're more able to cope with less than ideal soil conditions.

Enough money. If money is not painfully short, I'd improve the soil for low-demand vegetables as though I were preparing for growing medium-demand

These recommendations are minimums for growing low-, medium-, and high-demand vegetables on all soil types except heavy clays. Excessive liming can be harmful to soil. Do not use more than I recommend. COF is potent! Use no more than recommended here. However, it is always wise to exceed the amounts of manure and compost I suggest in this table by half again or double if you can afford to. No matter how much organic matter you may have available, do not apply more than double the amount recommended here or you'll risk unbalancing your soil's mineral content. If you think your vegetables aren't growing well enough, do not remedy that with more manure or compost; fix it with COF.

Steer manure: Sacked steer manure is commonly heaped in front of supermarkets in springtime at a relatively low price per bag. However, this material may contain semi-decomposed sawdust and usually has little fertilizing value. It does feed soil microbes and improves soil structure, which helps roots breathe. And it has been at least partially composted; it is not raw manure. It is useful if not hugely overapplied.

Chicken manure: Chicken manure in sacks (which has been somewhat composted but is not labeled "compost") or "chicken manure compost" is far better stuff than weak steer manure for *fertilizing* the veggie patch, but take care not to overuse it. The product I used to buy before I switched exclusively to COF was labeled with an NPK of 4-3-2, potent stuff. ■

ones. I'd prepare the soil for medium- and high-demand vegetables as though I were growing high-demand ones, and I'd also use COF instead of chicken manure compost. That's what I do in my own garden.

Clay soils

The suggestions above apply if you've got a fairly workable soil — lots of sand or silt and not too much clay (for a discussion of soil content, see Chapter 6). If you have clay, though, you'll have to do some extra soil preparation.

Clay soils do have some agricultural uses — if they drain well and aren't the heaviest, most airless sorts of "gumbo" or "adobe," they can be good for orchards and for permanent pastures. But no sensible farmer or market gardener would willingly use a clay soil to grow any crop that required plowing to make a seedbed — which is what veggie gardening is all about. Clay is incredibly difficult to work. If you dig it when it is even slightly too wet, it instantly forms rock-hard clods. It also gets rock-hard when dry, and if you have enough mechanical force at your disposal to plow or rototill it when too dry, it then forms dust. The first time this dust is rained on or irrigated, it slumps into an airless goo in which almost nothing will grow. There is only a short period during the drying down of a clay soil that it will form something resembling a seedbed when tilled. This fact means a clay garden can be mighty late to start in a wet spring. In Chapter 3 I describe a "ready-to-till" test you can use to determine when soil is at the right moisture to work. This test is essential when working clay.

Clay is heavy. Shovelful for shovelful it weighs twice as much as loam, so working clay soil exhausts the gardener. Even after it is tilled successfully (at the right moisture content), clay quickly settles into a hard mass, which makes it much more effort to hoe. As a result, it's not easy to get weeds out of a clay garden. To top it off, clay tends to contain little air, so most vegetables don't grow well in it.

Clay can be turned into something resembling a lighter soil by blending enough organic matter into it. Not the essential little bit that enlivens a soil containing less than half clay, but huge amounts, a layer several inches thick. But remedying clay this way is not wise for several reasons. First, because in the long run this practice is expensive. The high level of organic matter you must create to improve the structure of clay decomposes rapidly and requires replacement every year. The first year might take spading in a layer three to four inches (7.5 to 10 centimeters) thick to significantly open up a clay. In subsequent years you might only need to dig in a 1½ inch-thick (4-centimeter-thick) layer to maintain the benefit. That effort demands a lot of hauling. And spreading. And consumes your time spent making compost. And when all's done, the remedial clay will never grow vegetables as well as a light soil will — never.

I am well aware of what garden magazines (and books) have said about the joys of turning any old clay pit into a Garden of Eatin' by hauling and blending in enough composted organic matter. But please consider what I say in this book about the harmful nutritional consequences of unbalancing a soil after mixing too much organic matter into it. The quality of your garden's nutritional output *is* your family's health.

Remedying clay on the cheap. If your garden must be on a clay soil, and if you lack funds to do anything else and live where the native vegetation is a lush forest, my advice is this: before digging a new garden site for the first time, spread a layer of compost or well-rotted manure about one inch thick (2.5 centimeters) and spread 100 pounds of agricultural lime per 1,000 square feet (50 kilograms per 100 square meters). Add these large amounts only for the first year. If you live in a dry climate providing under 30 inches (75 centimeters) of rainfall a year (where soils may already contain significant levels of calcium and sometimes excessive magnesium, too) and have clay to deal with, I suggest that you consult with a local soil-testing service or agricultural agency before adding any forms of lime in excess of the minimal amounts in COF. You might be advised to use gypsum instead of lime.

A one-inch (2.5-centimeter) layer of decomposed organic matter blended into a shovel's depth of clay is not enough to make it seem light and fluffy, but it is enough to make a decent garden. You'll never grow prize-winning carrots, parsnips, or sweet potatoes in this soil, and it will never produce gigantic broccoli, but it will serve. In subsequent years you can manage your garden as I suggest for other soil types, but take extra care not to dig it when wet. And especially when working clay, if you want to end up with a reasonable seedbed after the first year, put your amendments of compost or manure into the surface with a rake after digging, as will be described in Chapter 3.

Chemically, clays act like discharged lead-acid batteries. They have the capacity to soak up huge quantities of plant nutrients. Until they have been brought close to "full charge," they tend to withhold nutrients from your vegetables. I strongly suggest that gardeners with clay make and use COF, applying it at about 1½ times the usual rate for the first few years.

Remedying clay when you're able to invest. Almost no one who considers a vegetable garden one of the most important things in life and has had the experience of trying to garden on clay would ever buy a home surrounded by clay soil unless there were some other compelling reasons. But if you are already located on clay, there is something you can do that is a lot cheaper than moving, something that will result in a garden nearly as good as one situated on a much lighter soil. Simply import a thick layer of loam topsoil and spread it above the clay.

I confess: I did this for my current garden. I live in an absolutely beautiful but simple home surrounded by six acres (2.5 hectares) of healthy native forest full of animal and bird life. It is on a steep sun-facing hillside with long views of a large river below us and of the mountains beyond seen through the trees. I love living here in the bush: I have no useless lawn to mow to please Everybody Else, and in the mornings I see numerous kangaroos grazing just outside my windows.

On my entire property there is only one area that is relatively flat and sunny enough to make a garden. And the soil there is a deep black clay. So I had many truckloads of topsoil dumped in a huge heap right in the center of the new garden site. I bought enough soil to cover all my beds (not the paths) nearly one foot (30 centimeters) thick. Next I erected a wildlife fence and then, with a wheelbarrow, began spreading that topsoil, creating one bed after another. What I ended up with is a garden where the topsoil is an easy-to-work loam, and what had been a clay topsoil became the subsoil.

I might mention that before covering the beds with topsoil, I limed, lightly manured, and rototilled the clay, thus accelerating the improvement of my subsoil by several years. Had I not done this laborious task, the clay would have loosened up anyway, but far more slowly. Eventually the worms would have transported humus into the clay, and leaching would have transported mineral nutrients into it. Eventually.

It cost me approximately $1,200 to have 120 cubic yards (100 cubic meters) of sandy loam topsoil delivered — a small investment, actually, when you consider the value of what I can grow with it. However, over the next few decades this investment will work out to be the least costly of all alternatives, and one that will pay large dividends for as long as I use the garden. Instead of being on the treadmill, having to haul and spread 20 to 25 cubic yards (15 to 20 cubic meters) of compost the first year to "build up" my clay into an inadequate imitation loam, and then another 10 to 15 or more yards (8 to 12 meters) every following year to maintain it in that condition, I now need only two or three cubic yards (1.5 to 2 cubic meters) a year to enliven my genuine loam topsoil. Because my soil has not been given too much organic matter, the nutritional qualities of my vegetables are the highest they can be. Because I am digging a light soil, I can work my garden early in spring when it is wet. Because the soil is light, I don't get tired from digging it and it forms a fine seedbed with little effort. The only crop in my entire garden that does not grow as well as I might hope is celery, which, to grow really well, requires a light loam soil at least three feet (one meter) deep.

Soil temperature and oxygen

Here's something you probably don't already know about plants. The business of construction, of growth, is mainly done at night. Plants store up energy by converting sunshine, water, and air into sugar during the daytime. They burn that sugar as energy to grow with during the night.

Assuming the plant lacks for nothing in the way of nutrition, moisture, etc., then the speed with which a plant grows is determined by temperature. For every 10°F (5°C) increase in temperature, the speed of growth doubles, meaning the size increases from 1 to 2, then to 4, and then 8, 16, 32, etc. When the nighttime temperature is much below 50°F (10°C), all biological chemical reactions, including growing, go on slowly; nothing much seems to happen. Even fruit won't ripen much when the nights are chilly because ripening is just another form of growing. Let's say that during a chilly evening when the temperature has dropped to 50°F at sundown, the plant will grow one size larger overnight. If the nighttime low is 60°F (15°C), the plant can grow two sizes larger. If the night is a warmer one, when the low is 70°F (21°C), the plant enlarges by four sizes overnight. On a steamy evening when people can hardly sleep and it is 80°F (26°C) at sunup, all other things being equal, the plants in the garden may have grown eight sizes overnight. That's why on a hot midsummer's night you can hear the corn growing.

This principle also applies to sprouting seeds, which is another growth process. Some kinds of vegetable seeds (spinach, mustard, radish) can sprout in chilly soils, though slowly, usually taking two weeks to emerge, where in warm earth they would be up and growing in four or five days. Others (melons, cucumbers, okra) need warm soils to germinate at all and either sprout quickly in warmth or rot in the earth if it is too chilly.

One of the most educational bits of equipment new gardeners can have is a soil thermometer. Watching it closely over a few springs allows them to grasp the connection between the daily weather, the rise and fall of soil temperature a few inches below the surface, and the resulting rate of seed sprouting and plant growth. It is not only air temperature and the amount of sunshine that determines how fast a plant grows. For the tops to grow, the roots have to grow, too. And, all things being equal, the speed at which roots grow is determined by the soil's temperature.

TV documentaries and botany books all tell us that when plants use solar energy to make sugar, they take carbon dioxide from the air and break it down into oxygen (which they can breathe or release into the air) and carbon, which they combine with water into sugar. This is true. What books and movies don't point out, though, is that, without sunlight, the plant can't do photosynthesis and so has to breathe oxygen like any other living thing.

The roots do not conduct photosynthesis. And oxygen can't be sent from the leaves to the roots. So the roots, being alive, need to breathe in oxygen all the time, and like other carbon burners on our planet, they breathe out carbon dioxide gas. How do they get oxygen? At a casual look, the soil might seem like a solid thing, but it is actually made of little particles that have spaces between them filled with either water or air. The larger the air-filled spaces are, the more freely air can move between the particles and the more freely the soil air, which gets thick with carbon dioxide exhaled by all the living things in it (including plant roots, soil animals, and microorganisms), can be exchanged with atmospheric air, which contains the normal amount of oxygen. The more oxygen the roots can get, the faster they can grow. Growth above ground matches what is happening below ground. If the roots are choked for lack of soil oxygen, even if there is otherwise plenty of fertility in the soil, the growth of the top will also be reduced.

One reason we dig before planting is to get more air into the soil, to fluff the earth up and temporarily create more air spaces in it. But within four to six weeks the soil will settle back to its natural degree of compaction. So plants grow faster for a month or so after spading, but then their growth slows down. Most vegetable crops occupy their beds for 12 or more weeks, and a few will grow on for five months or so, but you can't spade up a bed after the plants are in and growing. There is a remedy: Digging in compost or manure loosens the texture of the soil far better, and for far longer, than digging alone will do. This better texture is called *tilth,* and the improvement lasts for a year or three, depending on how much manure/compost was dug in and how fast it rots away.

Summary

This chapter has provided the bare minimum of information and some basic techniques. If you knew nothing more than this, and if you spread manure and lime (or, better, manure and COF), dug the ground once a year, followed the vague instructions on the back of seed packets, put in some seedlings and a few seeds, hoed the weeds, and did a bit of thinning, you would have a productive garden. But there is quite a bit more to be known.

The next chapter will explain how you can use hand tools to work the soil more effectively, show you a better way to manure the ground in an ongoing garden so that your seeds will germinate more effectively, and describe a number of other techniques that will both save you time and make your efforts pay off better.

CHAPTER 3

Tools and tasks

This chapter is about making a garden without expensive gear and about how gardening with manual labor is neither exhausting nor difficult. It is rather easy, however, to fool a novice into believing he or she needs a powerful tiller. If you purchase any kind of shovel or spade and an ordinary hoe and try working some hard soil with these tools exactly as they came from the store, digging will instantly seem the most exhausting thing a person ever tried to do. You will conclude that anyone who can make a ditch with a shovel has to be the strongest man that ever walked the earth. Try chopping weeds with that new hoe and after two minutes you'll be dripping buckets of sweat, tired out for the rest of the day.

The reason: When you bring them home from the store, *the tools are never sharp.* Worse, most gardeners never properly sharpen their tools after buying them. I know this for a fact. Once a year, in spring, I teach a veggie gardening course, and for one session I ask everyone to bring in their hoes and shovels and we sharpen them. This turns out to be the second-most-popular thing that happens in my class. (The event participants think is the most useful is the morning spent in my own garden, when I demonstrate how to prepare a raised bed with those tools and how to sow seeds so they'll come up.)

The basic three and a file

To handle a garden up to a quarter acre (1,000 square meters) in size, if you are minimally fit (I said "minimally fit"; I did not say male, nor did I say huge), only four tools are essential: an ordinary combination shovel, a common hoe, a bow rake, and a 10- to 12-inch-long (25- to 30-centimeter) metal file (with handle) to periodically sharpen the shovel and hoe. If you are not fit, then you'll probably want to get someone else to work the earth for you the first time it is dug (or have it rototilled). If you are physically unable to shovel or hoe, you'll probably have to consider the second-rate practice of mulch gardening. I do not recommend permanent mulching unless, because of infirmity, there is no choice.

Please do not buy cheap discount store tools. If you are what Australians call "skint," (broke), you will do better pawing through secondhand shops until you find good ones. Where can you find quality new tools? I suggest a visit to a commercial hardware store, landscaper's or nursery supply company, or contractor's supply store and inspect what they sell to tradespeople. People working with such tools all day long can't afford to be resharpening one every half hour, breaking one every other day, or wearing them out every few months. Quality tools aren't cheap, but they work out to be the least costly in the long run.

On the other hand, in some mail-order catalogs you'll find tools, supposedly of extraordinary quality, offered at extraordinary prices. These status symbols are usually no more effective or long-lasting than the ordinary tradespersons' gear.

The shovel

There are different kinds of shovels and spades, each designed for a different task. Industrial hardware stores and landscapers' have makers' catalogs on hand showing each design, weight, and size. I suggest you ask to study the catalog — and take your time with this.

Some **shovels** are designed to move piles of loose material from one place to another — for example, to fill a wheelbarrow with sand or small gravel that is spooned up from a heap of loose material. Blades on these kinds of shovels are broad and deeply curved so as to hold a lot, like a soup spoon. This tool is designed to be thrust into a heap of soft material with the strength of the arms and upper body and then to lift out a large measure, so the blade rarely has a rolled-over top (to cushion the foot pressing it into the earth). It is not easy to dig into the earth with one of these.

The spade, on the other hand, is designed to dig straight down efficiently and loosen compacted soil. Usually the foot-long (30-centimeter) flat blade is only six to eight inches (15 to 20 centimeters) wide, so it doesn't take too much weight to push it into the earth. I've seen this sort of shovel made with a short blade, as though for children, but I would never buy one with a blade less than ten inches (25 centimeters) long; a foot long is best. The blade is pressed straight down into the earth by one foot, with the gardener applying the whole weight of the body, so the top edge is rolled over to avoid bruising the foot pressing it into the earth. Spades are marvelous tools for loosening soil; of all digging implements they will penetrate harder earth with the least force or weight atop them. The limitation with a spade is that because the blade is flat and narrow, it is inefficient when it comes to turning over or moving the earth. Spades are best for working compacted new ground for the first time, but after that they are entirely outclassed by the ordinary combination shovel (see below).

I own a special long-handled spade, locally called a "plumber's friend." This one has a slightly pointed cutting edge on a rather narrow and strong slightly curved blade that is as effective at penetrating compacted earth as an ordinary spade, but will move soil better than a flat blade. It is ideal when you're trying to unearth a leaking pipe. I use it only to break new ground. Most of the time it sits unused in the garden shed. But when I have occasion to use it, the plumber's friend is my friend, too, because it saves me a lot of work compared to doing that job with a combination shovel.

The combination shovel, as its name implies, does what both the spade and the shovel do, but in one tool. The blade is a bit wider than a spade's, but still narrow enough that it doesn't take an unreasonable amount of weight to press it into uncompacted earth. The blade is curved enough that it can pick up loose material and toss it somewhere. Because its design is a compromise — allowing it to dig firm but not rock-hard soil and toss loose stuff — it is not ideally efficient at either task. When you are buying a combination shovel, make sure the top of the blade is rolled over to facilitate pressing down with the sole of the shoe. It should be a long-handled tool so that if the soil is a bit resistant to penetration, you can wiggle the handle as you put most of your body's weight on the blade, standing on it balanced on one foot. Once the blade has been inserted, the curve of the shovel allows you to lift a fair amount of earth and easily flip it over. But if you should have a wheelbarrow full of compost or manure, the combination shovel also allows you to lift that loose stuff and spread it by tossing it considerable distances. You can't do this effectively with a spade.

Short-handled shovels and spades are for people under five feet (1.5 meters) tall. Bending over while working a shovel or spade is exhausting and soon becomes painful unless your back is extremely strong. Short-handled tools are good to take camping or to toss in the trunk to use in an emergency. And maybe they're good for children. But for regular garden use, there is no way I would ever have one.

Spading forks are designed to act like a spade, but they also rapidly break up soil when they are levered back after being inserted because the earth can pass between the tines. They are easier to push into the earth than a spade. Spading forks work effectively only on naturally loose (sandy) soil. The tool is worse than a spade at picking up soil to flip it over. And the tines will bend when used with enough force to work fairly heavy ground.

In my opinion short-handled spading forks are an instrument of torture, but I am nearly six feet (1.8 meters) tall.

All things being equal, I recommend the combination shovel.

Sharpening the shovel

It is highly unlikely that any shovel (or hoe) will be sharp when you purchase it. In fact, it will almost certainly be entirely blunt — squared off instead of honed into an acute bevel like a chisel's edge. Sadly, as is also the case with knives and a lot of other tools that have to be sharp to be effective, the only way you can find out if you purchased a good piece of steel is to sharpen it, use it, and find out how long it stays sharp. One indicator: the slower the stone or file cuts the metal, the better the steel probably is.

Making tool steel inevitably involves compromise. The harder the steel is, the sharper it'll become and the longer it'll stay sharp. Usually. However, the harder the steel is, the more brittle it is and the more likely the edge will chip or the blade will snap when twisted or bent. Old-fashioned straight razors are extremely brittle, made of the hardest possible steel. Axes smack into things with a great deal of force; they must be made of much softer metal or else they'd shatter when hitting a knot. Garden implements may encounter stones and be on the receiving end of a lot of bending stress; they need an intermediate grade of tool-quality steel.

In my opinion, making something of cheap material that only resembles a working tool is dishonest and takes advantage of ignorance. Tool steel is costly to buy and costly to work. I know of no way to tell with reasonable certainty if I'm about to buy a good shovel or a piece of junk, other than to select a brand name used by tradespeople. I also look for a few indications. Is the handle solidly attached; is the tool light and well balanced, yet strong. If it is a combination shovel or spade, is the top of the blade rolled over so you can press hard with your foot without pain. And, of course, you can attempt to sharpen it. If the stone or file cuts the metal rapidly with little pressure applied, it is soft metal that will rapidly get dull with use.

To sharpen a shovel, place it on the edge of a porch or on the earth. Put one foot on the handle, just behind where it sockets into the blade, and press down with enough weight to hold it still while you file it, or do it the easy way — have someone else hold it for you. Then, while leaning over the blade, hold the file so that the angle between the blade and the file is about 15°. Maintain that angle firmly. *Do not let the file change its angle.* Now stroke the file along the edge, pressing down firmly but not so hard that you lose control of the unchanging angle of the file. If the cutting edge is curved, always start the stroke in the center of the blade and work out toward one side. Do a few strokes from center to one side and then do a few strokes going to the other side from the center. After a few hundred strokes (it may seem an endless task the first time you sharpen a new tool), you will have removed enough metal to produce a chisel-like bevel, and the shovel will have a sharp edge, nearly sharp enough to cut your finger on. If you want the edge to be really sharp, the bevel must have no variations in it, no changes of angle, and the entire cutting edge must have the exact same angle of bevel.

It is not necessary to make a bevel that goes all the way up the rounded sides because when the shovel is cutting its way into the earth, most of the effort of cutting is done within four inches (ten centimeters) of each side of the center.

As the bevel is being created, a "wire" will form on the opposite side. The action of filing forms a thin sheet of metal that wraps around the edge of the blade. This wire prevents the edge from being really sharp and should be

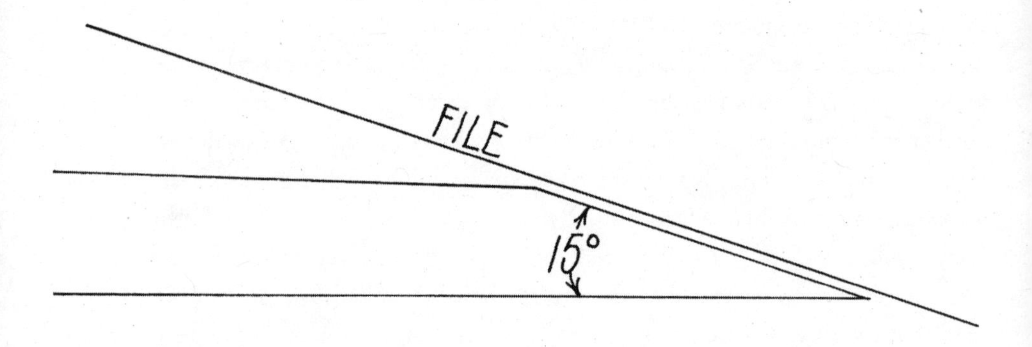

Figure 3.1: *The shovel's bevelled edge. Hoes should not be ground to an angle this acute; 20° is correct for them.*

removed. It is easy and quick to do this. Just flip the shovel over and, holding the file at about a 5° angle, make a few light strokes. Not many, just a few. You want to remove the wire, not make a matching bevel on the back more acute than the one on the front.

A quality shovel will dig all summer and not get dull unless you blunt the edge striking a good many rocks. It does not take an impossible amount of effort or weight to push a sharp combination shovel or spade into the soil. A fit woman weighing over 100 pounds (45 kilograms) should be able to manage it just fine. Someone weighing even less can do it in soil that has been spaded up once and has then received small yearly additions of manure or compost, which keeps the soil loose and friable. Someone who weighs under a hundred pounds can still push in a spade that is a bit narrower than the usual.

If I were shopping for a second-hand shovel I'd take my own file to the Salvation Army thrift store and tell the clerk when I came in that this was *my* file and I had brought it to test some shovels (and hoes). Then I'd make a few strokes on the edge of any likely-looking tool. The speed at which the file cut would reveal everything about the tool except how strongly the handle was attached. Cuts fast: junk. Cuts slow: probably good steel. Stroke a few with a new sharp file and you'll instantly see the differences.

If you don't abuse it, a shovel should be a long-lasting tool, but even quality shovels wear out. How fast they wear out depends on how much they are

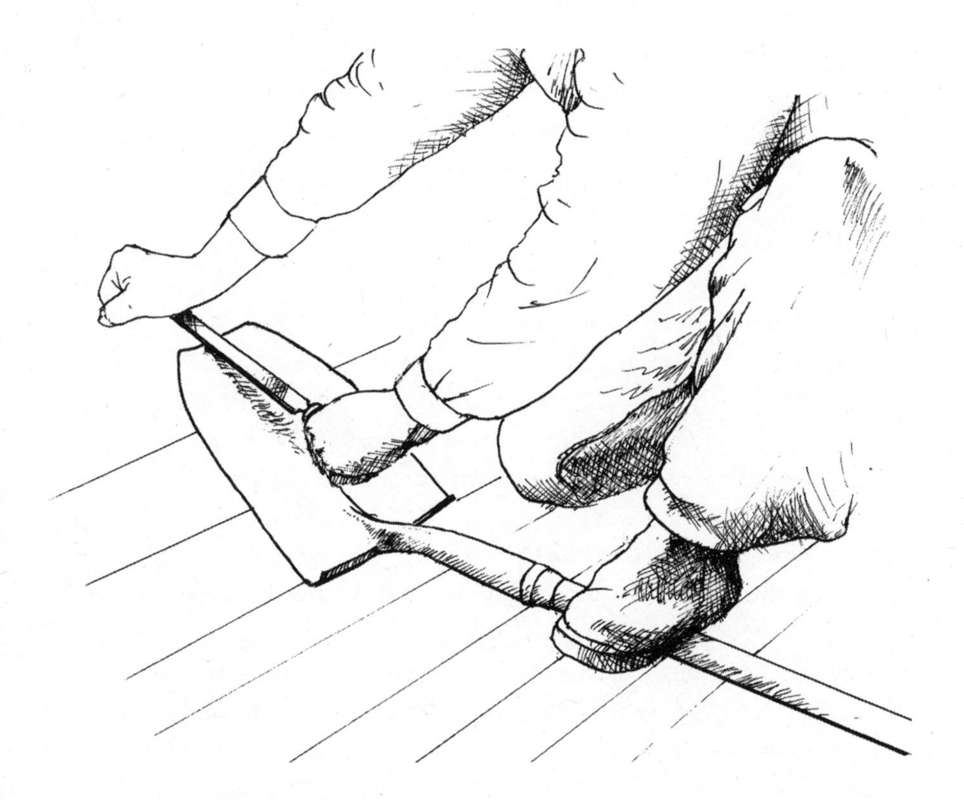

Figure 3.2: *Sharpening a shovel.*

used and the nature of the soil they are being used in. If the soil is sandy, then repeatedly pushing the blade into that soil is a bit like rubbing it with sandpaper. The blade gets steadily thinner (and weaker) and eventually will break. (The same thing happens to rototiller tines; they'll wear down to nubbins.) Expected life of a shovel that digs a 2,000-square-foot (200-square-meter) garden several times every year: 10 to 20 years.

If I were shopping for a used shovel, I'd inspect the blade closely for fine cracks, an indication that the steel has worn too thin. The presence of those cracks indicates the shovel's useful life is over.

The file

Files are made of extremely hard steel, but do get dull with use. Once a file has dulled (or rusted), it takes forever to remove much metal with it. Start out with a new one, ten inches (25 centimeters) long with a solid handle attached. When filing seems to need more time than it used to take, get another file. They don't cost much. And keep a light coat of oil on yours if it'll be sitting around unused for months on end.

How to start a new garden

New gardens are almost always made in a place that had been growing grasses, often mixed with other low-growing plants. Grass makes an especially dense network of tough roots, called sod, that holds soil together firmly. If the grass grew thickly, the soil will not immediately break up after hand-digging. It won't break up until the sod rots. And grass is tough, a survivor. It won't die and rot just because it has been turned over once; it'll attempt to regrow, so you've got to work to kill it.

In the Deep South of the United States, in most of California, and Down Under, this task can be done at any time of year that the earth is at the right moisture for digging. In harsher climates, the best season for this task is not spring, when many acquire a passion for gardening, but at the end of summer in anticipation of next spring. Whichever season it turns out to be for you, first mow the grass as close to the ground as possible — scalp it right down to the growing points if doing so won't damage your mowing gear. Mow until you expose patches of nearly bare soil.

If the clippings are short, let them rest in place. If it was such tall grass that you had to cut it with a weed whacker, scythe, or sickle, it would be a good idea to rake the grass up and remove it to make compost; your shovel won't cut through it easily. You also don't want to be paying the hourly rental on a walk-behind rotary cultivator that repeatedly gets its tines tangled up.

This done, spread the basic amendments to enrich your soil the first time — lime(s), complete organic fertilizer, manure and/or compost — in the amounts recommended in Chapter 2.

No matter which way you choose to do the job, rototill or hand-dig, or when, summer's end, autumn, or spring, do it only when the soil moisture is right. If the soil is too dry, the ground will be so hard that the task will take at least three times as long. Wait for rain or irrigate first. If you dig wet soil that contains much clay, you'll likely make a lot of hard lumps, called clods. When these clods dry out, they become almost rock-like and won't break down for months, maybe not until they've gone through the next winter. It can be extremely difficult to get seeds to come up in cloddy soil, and plants won't grow as well in it either. If it is spring, wait until the soil has dried enough to work. If it is autumn, making a few clods is not such a problem as they'll break down over the winter.

The "ready-to-till test." Take a small handful of soil and mold it as firmly as you can into a round ball about the size of a golf ball. Pack it hard. Then, cradling that ball in your palm, press firmly on it with your thumb. If it breaks apart easily and crumbles, the soil is just right to work. If the ball won't crumble, and your thumb merely makes a dent in the ball's gooey side, it is too wet. If the soil would not form a ball that sticks together, it is either too dry or contains no clay. A clayless soil may feel damp in your hand but will not form a ball no matter how firmly you mold it. That means you have a soil that won't form clods no matter what you do, so feel free to work this soil at any time, wet or dry. (You also have a soil that won't hold much moisture, and unless there is a second layer of soil located not too far below the surface that will hold a lot of moisture, this site is going to make a droughty garden that probably won't do well unless you can regularly irrigate it.)

Killing sod by rotary cultivation

The fastest way to kill a grass sod without using herbicides is with a rototiller. In one afternoon you can change an area of sod into something resembling a ready-to-plant seedbed, with fluffy loose soil that extends down about five inches (12 centimeters). I've had much experience with tilling. I have hired men owning large tractors do it for me by the hour; I did my variety trials ground myself using a self-propelled seven-horsepower rototiller. Before that I owned a "front-end" tiller. Now that I am old and weak (and wiser), I prefer digging with a shovel, a somewhat slower approach that requires no gas, makes no noise, and actually takes little more effort (but is spread over more time) than tilling does.

Rototilling is pleasing when you are in a hurry. Even if you're flush with money, I suggest, when starting a new garden, that you hire a machine or hire someone who owns one to come in and only rototill on that one occasion. The last thing you want to do is own one because you'll get far better results using a shovel.

Do not try to eliminate grass/sod with a "front-end" rototiller, which will prove several times more exhausting (and little faster) than doing it by hand. Use only a self-propelled rear-tined tiller of at least seven or eight horsepower, or hire someone. To estimate what tiller hire will cost, an effective walk-behind rear-end tiller can eliminate about 1,000 square feet (90 square meters) of grass per hour, digging to about five inches (12 centimeters). If the soil is dry or particularly hard, or if the sod is particularly tough, you might double that estimate. It would take me about 20 hours to dig that much new land *by hand the first time,* but I would do it to 12 inches (30 centimeters) deep. If you sprayed that area first with the herbicide glyphosate and allowed enough time to pass for the sod to die and most of it to rot (four to eight weeks of warm weather, or over winter if you spray in early autumn), the digging or tilling would go at least twice as fast. (This assumes you are not gardening strictly organically.)

Here are a few hints for tilling. My instructions may seem unclear, but, believe me, this information will make sense as soon as you actually start.

I am assuming the plot is a rectangle. Make the first pass straight down the center the long way. Why the long way? Because turning around is an effort and wastes time; the less often you have to do it, the better. Try hard to hold the first pass exactly straight. There will be a drag bar that controls how deeply the tines cut. Set it so you dig only one inch (2.5 centimeters) below the surface; otherwise the machine will tend to skip and jump in all directions. Do not rent a walk-behind tiller that doesn't have a functioning drag bar.

Okay, you've done one pass the long way. Unless this particular soil and sod is softer and more tender than any I've ever met, it will seem to you that almost nothing has happened. Turn around and make another pass going the opposite direction, exactly down that same row you've already tilled. After you have made two or four or five such passes, each right on top of the previous ones, you'll start to see some soil; the grass will seem damaged. If you don't see soil, set the drag bar a step deeper.

When you start seeing soil, move over half the width of the tines and do another pass. The tiller will tilt to the side, leaning into the area you have tilled already. That is good; it'll do some more damage there and begin to damage a new half-row of sod. Somehow keep it going straight! At this point you might also try raising the drag bar another notch and see if the machine is still controllable when you try to cut deeper. If you've never done it before, tilling sod will be a lot more work than you expected, based on the pictures in garden magazines showing smiling people strolling along behind a tiller that is cruising through soft, recently tilled soil.

When you get to the end of the row, swing around and start down the other side of the first row you made. Your tiller will now tilt into the other half of the first row. Keep to that pattern, slowly widening the tilled area. If the plot is more than 50 feet (15 meters) wide, you will soon see it would have been cheaper to hire someone with a powerful tractor pulling a massively heavy four-foot-wide tiller.

This is important: If you're using a walk-behind tiller, do not expect to till deeply at first. If you work the whole area only two inches deep the first time over, you've done well! The result of that will be to kill almost all of the grass. The wisest thing to do at this point is to quit. Let your tingling hands and throbbing ears have a break. Let the sun and the passage of a few days help you kill off the grass you so valiantly damaged. If you should till that plot deeper now, you'll likely relocate some fragments of the grass into soil that'll stay moist enough for it to start growing again. My advice is to take the tiller back to the rental yard and come back in three to seven days and attack the plot again, this time with a spade or combination shovel.

However, I know many of you are not going to take my advice. So be it. In that case, the second-wisest action is to cross-till. Make another set of passes over the ground, this time across the ones you made first. If you tilled east to west the first time, now do it north to south. Use the same pattern; start in the center and work outward to the edges. If the soil seems a bit softer now that you are below the thickest of the grass roots, lift the drag bar a notch and see if you can't get down two or three inches (five to eight centimeters) farther on these cross-passes. This time, with luck and a powerful machine in front of you, you'll end up maybe four to five inches (10 to 13 centimeters) deep.

You should quit here; you'll be tired for sure. But unless your thin wallet is counting every second of rental time, I know you'll want to go on and try to make a seedbed out of the entire garden. If so, cross-till again. Go back to the direction you made your first passes in (the long way) and do it all over again. This time lift the drag bar another inch or so and go for it ... as deep as you can. If the tiller has a handle that will swing from side to side, offset it a bit so your footprints are on the ground that will be tilled the next pass after the one you are doing now. With a bit of luck you'll proudly end up with a fluffy-looking patch of soil that contains only a few footprints.

I want you to notice those footprints. They should have a lesson to teach you. Immediately after you till, all the soil will be damp. But after the sun dries out the surface, it will change to a lighter color. Early the next morning, before the sun shines intensely on the soil, visit your plot and look at those footprints. If it hasn't rained, the whole tilled surface should still be dry except for the bottom of those footprints. They will be damp. Why? Because your weight pressed the soil particles together firmly enough to allow moisture from deeper layers to rise up. There is a term for this movement of water — capillarity flow. Understanding capillarity will help you sprout seeds more effectively and also grow a garden without irrigating it at all (or at least watering it far less frequently), so I'll have more to say about it in Chapters 5 and 6.

One more thing I want you to notice if you have elected to use a walk-behind rototiller. How deep did it actually go? If you managed to get the tines in as far as they would go, push your hand into this freshly fluffed-up earth and measure. Seems like eight inches (20 centimeters), doesn't it? Well, wait a few days for the soil to settle back down to a more normal compaction and check again. Now how deep did you till? If it was a walk-behind tiller, not more than five inches (12 centimeters). If it was a big tractor sporting a heavy rotary cultivator run by an experienced operator doing honest work, seven inches (18 centimeters). Maybe.

And that is the trouble with rototilling. Loosening soil six or seven inches deep is not enough. Vegetables do not like their root systems to be so confined. Even worse, if the soil has much clay in it, the bent-over ends of the rotating tines will have compressed a layer of soil located at the farthest depth they reached. Immediately after tilling, if you'll push away the just-tilled fluff over a spot, you'll see the shiny polished "plow pan" I'm referring to. The pan is more compacted than the original soil was before you started tilling. The plow pan acts as an effective barrier to root penetration and tends to limit the depth of most vegetables' root systems to the soil above it. This is not good!

To grow a really great garden you need loose soil to a depth of at least one foot (30 centimeters), and you do not need any major barriers in the way of your crops' roots penetrating several feet farther down. The only practical way to do this on a garden scale is with a spade or combination shovel, or with a spading fork if you've got light soil. And that's why a few paragraphs above I twice gently suggested — after the first tilling and then again after the first cross-tilling — that you quit tilling. At that point you have already done most of the good that can be done with rototilling: the grass is pretty much destroyed without resorting to glyphosate; and the hardest, most compacted soil of the top three inches (eight centimeters) is pretty well broken up.

From this point the rest of the job — working the earth to a foot deep and making seedbeds — can be done by hand with a lot less effort. After tilling the first time, you can treat the ground as though you had an established garden going, follow my suggestions for restoring beds, and do what I soon will suggest for making hills and wide rows for a new planting season. It'll all be a whole lot easier.

I do need to say one more thing about a plow pan. It is called that because a far thicker and longer-lasting compacted layer than a rototiller will make is routinely created by the common moldboard plow. The plow's bottom, or "sole," slides across the unplowed soil about seven inches (18 centimeters) down, resting heavily on the soil and pressing down at the same depth every single time the field is turned over. Repeat plowing over a few dozen years creates a hard layer several inches thick that starts seven inches below the surface. This layer forms an effective barrier to root penetration. The consequence is that the crop tends to grow in the top seven inches of the field. The existence of more soil below the pan is of little use to the crop. Take a look at some of the drawings of root systems in Chapter 9 and try to imagine how poorly a vegetable will grow if it ideally has a root system four feet (120 centimeters) deep, but is largely restricted to seven inches (18 centimeters). (For more on this, I suggest you read *Plowman's Folly*, listed in the Bibliography.)

Why am I going on about plow pans in a garden book? Because if your garden is on land that once was a farm field, it is likely to have a plow pan. That compacted layer will not go away simply because plowing has ceased. A plow pan may persist for half a century. The only way to get rid of it for a garden is to dig through it with a shovel or fork.

Even if you have created the illusion of loose soil with a rototiller, there is the tiller "pan" and maybe a plow pan, so you still need to deeply hand-dig the rows and beds. However, once you have eliminated the sod, hand-digging goes much easier.

Killing sod by hand-digging

After you've mowed the plot closely (and removed the mowings for composting if they make a thick mulch) and spread soil amendments, as described at the beginning of this section, you start to eliminate the sod. Basically you dig up chunks of earth and turn them on their side or, better, upside down (if the chunks are not too big and you have the strength to do so). This damages the grass and will cause quite a bit of it to die. Then, about a week later, just before the sod starts growing again, turn it over once more, cutting through the chunks while making the second turn, to reduce their size. This second turn will finally kill most of the grass and blend in the amendments more effectively, and you'll also dig deeper, quicker, and with much less effort. On the third turn, after yet another week passes, enough of the sod will have decomposed that you'll be well on your way to having fine, crumbly soil nearly to the depth of the blade. Any preexisting plow pan will be gone.

Here are some tips. While spading, the bigger the chunks that you try to bite off, the harder you'll work. Be methodical and precise; nibble off bits of a size and weight you can comfortably handle. Large tasks like making a new garden can seem too daunting if you contemplate doing the whole thing at once, so I suggest that you divide your garden into beds of 100 square feet (ten square meters) and dig one bed (and one between-bed path) at a time. Using corner pegs, mark out strips about 5 feet (1.5 meters) wide (which is the length of an ordinary shovel, tip of handle to tip of blade) and 25 feet (eight meters) long. Start out standing at one end of that strip. Put the shovel's blade only one inch (2.5 centimeters) back from the end of the strip and push it in as far as you can manage. If the sod is strong, this first nibble probably will go in only an inch or two. Lever out this bit of soil and roots and put it into a wheelbarrow. Then move the blade back only an inch or so and push it in again. This time it should go in a bit deeper. Lever out that bit of soil and put it in the wheelbarrow. Repeat. By the time you have worked back five or six inches (12 to 15 centimeters) from the end of the bed and across its full five foot width, there will be a small ditch and you will probably be able to insert the blade to half its length. And the wheelbarrow should be full.

Continue digging back and forth across the bed the five-foot way, lifting small chunks and placing them on their sides or upside down in the depression you made while nibbling out the previous row of chunks. The softer the earth naturally is and the bigger the digger, the larger the chunks you can move and the deeper you can go. But big or small, deep or shallow, even if you're only moving back an inch or so with each effort, gradually the whole bed will have been broken into chunks six or seven inches (15 to 18 centimeters) deep and turned over. When you get to the end of the bed, put the contents of the wheelbarrow into the hole left from turning the last row.

If you dig to kill the sod in the spring, there will be a few weeks' lag between the first dig and the start of planting. If you start this preparation in the autumn (including incorporation of amendments), you'll be all ready to go in spring as soon as the ground can be worked. Another advantage of digging in autumn is that there's always the odd bit of grass that isn't killed. To prepare the autumn-dug beds for sowing in spring, you're going to dig yet again and this time you can yank out any resistant clump of still-living sod and toss it onto the compost heap. Yet one more advantage of autumn digging: A new garden is usually full of weed seeds. When you turn it over in late summer, a lot of those weeds sprout in autumn; the next spring, a lot more of them sprout. But before they have a chance to get large, you dig again and kill them off wholesale.

Turning over sod the first time is hard work. I am 63 years old, but I have been doing garden work for 33 years and consequently am reasonably fit. Still, I find that after turning new sod for about 90 minutes I am getting tired. And 90 minutes is about the time it takes me now to spade up an area of sod of about 5 feet by 25 feet (125 square feet or 11.5 square meters). So when I'm starting a new plot from sod, every day I dig one bed (and one path). After six days I will have dug 725 square feet (70 square meters) of new sod for the first time, a goodly area. And in consequence I am also in better shape than I was a week previously! I give myself one day of rest, and the next week, if there is more new ground to dig, I dig another 125 square feet each day. After digging that, I also re-dig the bed I first dug the previous week. This second digging doesn't take nearly as much work as the first time and goes deeper. I continue this pattern until the entire garden is dug up. Even if I were starting afresh with a fairly large plot that will contain 2,000 square feet (200 square meters) of growing beds and/or wide rows, I can have nearly the whole thing made into seedbeds in less than a month by working six days a week for about two hours a day.

If there is no sod, everything gets much easier. If I have recently done the area with a rototiller, even shallowly, then I can go back in with a shovel/spade/fork and in about 45 minutes break up 125 square feet to the full depth of the shovel's blade and not get particularly tired from the effort. Restoring an established bed or wide row for replanting only takes me about 30 minutes of moderate effort per 100 square feet (10 square meters). So why would I ever want to own a rototiller and experience the noise and vibration, use the gasoline, pay for the regular maintenance (or do it myself) on a touchy small engine, slowly march the tiller back into a shed when finished with it, etc.?

Raised beds and raised rows

By making a single choice, you can avoid much unnecessary garden work. You could remove weeds slowly and painstakingly by hand with dull fingers or rapidly and with little effort using a sharp hoe. Making small seeds come up in their growing beds could seem nearly impossible, forcing you to buy or raise transplants, or you could grow seeds sown directly in their final growing positions — easy as pie. You could spend hours getting a just-harvested area ready to replant or accomplish that task in minutes, with little effort. You could become a slave to a thirsty garden that needs watering after only one hot sunny day, or your garden could pretty much tend to its watering needs itself.

To garden the hard way, believe Everybody Else, who states that closely spaced, deeply and doubly dug raised beds, with vegetables painstakingly transplanted in precise hexagonal interplanted patterns, is the way to go. That authority may also tell you that you'll get a much higher yield, with the added bonus of putting less water in per unit of production out. None of these statements is exactly true.

As a novice I used that difficult intensive manner I just defamed. Gradually I learned that there are easier methods that allow you to harvest three quarters as much food (of higher quality) from a given area while doing a third as much work and irrigating a quarter as often — or not at all. I'm describing a system our ancestors used before every household had water under pressure coming out of pipes.

I now believe there is no one best way to arrange plantings. Raised beds are useful for some crops where and when there is irrigation water. Sprawling vine crops and sweet corn grow far better "on the flat" or in hills. Unirrigated gardens should mainly use hills and single rows except, perhaps, for small-sized vegetables.

Most intensivites say raised beds need to be double-dug to a depth of 24 inches (60 centimeters). I don't agree. Digging 12 inches (30 centimeters) deep is plenty; loosening the second foot produces little additional benefit in exchange for a heap of effort. Besides, over time the second foot will become looser without any extra effort on your part as worms transport the organic matter you put into the surface, and as plant nutrients leach into the subsoil (which chemically loosens clay — this phenomenon is called "flocculation").

It is far easier to get small seeds to sprout if the planting areas are raised slightly. Not because the seeds care about their position relative to the paths; they haven't a clue about anything except temperature, humidity, and the total frustration they must feel when a big heavy lump of impenetrable stuff sits above them. In that case, the seed dies. Large seeds like beans, peas, and corn can push harder against obstacles, and if you sow extra seeds, enough of them will usually emerge to make a row even in rough ground. However, tiny seeds like carrots, lettuce, and cabbage-family crops can't fight their way past a lump. They need what is termed a "fine seedbed" to come up in. After elevating the growing beds a few inches, you can make a finer seedbed with much less effort.

To understand what I mean, use a rake and try to make a zone of fine soil on flat ground after spading it up. After much time and effort you'll end up with a valley of fine soil, edged on both sides by ridges of clods trying to roll back down into it and smother your emerging seedlings. But if you first raise the beds a few inches, you can rapidly rake the clods and lumps out of the bed's surface and put them down into the paths, where your feet will break them up over the next months as you walk around the garden.

Note that you can only rake away the lumps if you have not put retaining walls around the outsides of the beds. Perimeter boards, railroad ties, elevated cinder block structures — all these attractive edgings eliminate the raised bed's two main advantages. The first big advantage — making a highly effective seed-sprouting medium — can be a piece of cake with a rake. Advantage two — keeping the weeds under control — can be done almost entirely with a sharp hoe, but you can't hoe right to the edges if the edges are made of something unhoeable. And grasses and other weeds that happen to come up along these retaining walls can get such a good grip that you will hardly be able to yank them out by hand.

Raising the growing areas a few inches also reminds people to keep their feet off. One reason raised beds grow better plants is that they tend to stay looser when feet aren't tromping on them. Walking on the root systems of growing vegetables harms them significantly and should be avoided. Walking on loosened ground before the vegetables put roots into that area compacts it and thus makes it much more difficult for the vegetables to develop roots in the first place.

Small-sized vegetables — beets, carrots, lettuce, bush beans, bush peas — grow best in raised beds or raised wide rows. Larger plants such as Swiss chard, broccoli, less-aggressive varieties of cucumber, cabbage, kale, and Brussels sprouts can also be grown atop raised beds and in wide raised single rows. The exceptions are all species that have large, vigorously sprouting seeds or that grow big, widely sprawling plants, such as squash, melons, and some kinds of

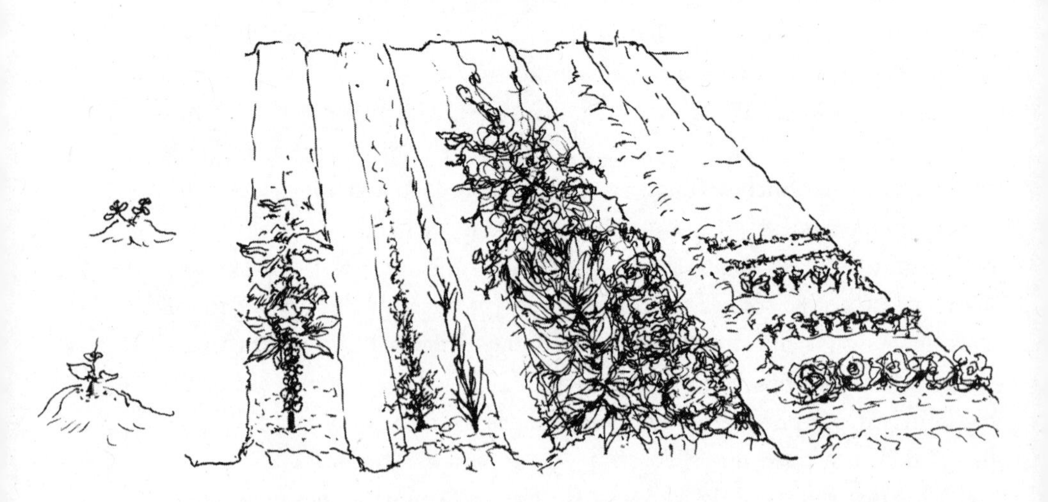

Figure 3.3: *From left to right, this illustration shows hills, two slightly raised single rows (the single wide raised row on the right has a double row of plants on it), and two raised beds. The raised bed on the left has two rows of large plants arranged on it, running lengthwise, and, farther down, a single row down the center. When doing this, a small, highly fertilized "hill" can be created in the raised bed, one for each plant. The raised bed on the far right has short rows arranged across it.*

tomatoes. Sweet corn, which needs to grow in a large cluster to pollinate, also does better "on the flat."

Making slightly raised beds

Raised beds should not be so wide that a person who is standing on the path next to them has trouble reaching to the center while bending over or squatting. For most people, four feet (120 centimeters) wide is about the limit. If you are inflexible or have short arms, the width can get down to as little as three feet (90 centimeters). But the narrower the beds become, the less efficiently the overall space will be used because there will be more paths. Also, if the bed gets much narrower than four feet, it is no longer possible to have two parallel rows of large plants running down it lengthwise. For someone who is quite flexible, with long arms, a bed might be 4½ feet (140 centimeters) wide. The paths between beds are about one foot (30 centimeters) wide.

Raised wide rows are better when gardening without any (or much) irrigation. These are actually just narrow raised beds, usually about two shovel-blades wide, or 20 to 24 inches (50 to 60 centimeters). There will be a one-foot-wide path between them, making the centers of the rows about 36 inches (90 centimeters) apart. The spacing between the rows should be increased if there may be long periods without rain. When I lived in western Oregon, where it almost never rains in summer, I used row centers of four feet and sometimes five feet (120 to 150 centimeters), depending on the crop.

Unless you have become so inflexible that you can no longer squat or bend over, there is no point in raising the beds or rows more than two to three inches (five to eight centimeters). This is easily accomplished.

Immediately after digging a new garden, using a combination shovel, move along what will be the paths, scoop up about two inches (five centimeters) of soil, and toss it atop what will be the growing bed or row. That's all there is to it. Stretching string lines or pegging the corners might help the first time these beds are made. In following years you simply redo the existing beds. Make wide raised rows the same way.

Making hills

Hills are best for growing pumpkins, winter squash, melons, tomato vines of an aggressively sprawling variety, and sometimes cucumber vines. I have also grown slightly smaller plants like kale, cabbage, Brussels sprouts, and broccoli in smaller hills made on top of a four-foot-wide (120-centimeter) raised bed, spaced 24 by 30 inches (60 by 75 centimeters) or even 24 by 24 inches. The traditional Native American spacing of hills is usually four feet from center to center but can increase to five or six feet (150 to 180 centimeters) when dry gardening where rainfall is low in the growing season. In the case of winter squash, hills could be as far apart as eight by ten feet (2.5 by 3 meters) because the plants grow so aggressively.

The best way to lay out part or all of a garden in hills is to first prepare the whole area for growing medium-demand plants. Spread lime, compost or manure, and COF, then dig the whole area. Then where the hills are to be, place a heaping cupful of COF or a pint or two (0.5 to 1 liter) of chicken manure and a heaping shovelful or two of low-potency manure or compost in a little pile. Then dig these amendments in, going down the full depth of the blade, working up a circular spot about 12 to 18 inches (30 to 45 centimeters) in diameter, thoroughly blending the amendments. You should end up with a little mound of highly fertile soil. If there isn't much of a mound when you're done, mark the spot by taking a bit of the earth surrounding the fertilized area and toss it on top of the hill. For smaller plants on raised beds, give the hills about half as much fertility and dig a smaller mound, maybe ten inches (25 centimeters) in diameter.

Plants started out in hills get off to a fast start because of the high fertility created immediately below where the seeds are sown. Hills are frugal; it would not be practical to try to turn the whole garden into soil that fertile. In Chapter 9, where I discuss how to grow various crops, I will suggest using hills when that method is appropriate.

Hills and survival gardening

Imagine this: It is spring planting time, but nothing has been dug and you urgently need to make a garden. There is sod. You have a shovel, some seeds, and not much else. What to do? Garden like the Native Americans taught us when we first arrived in North America. Don't even consider growing any demanding vegetables. You'll grow easy stuff: winter squash, corn, sunflowers, zucchini, melons, kale, large-framed late-maturing cabbages, indeterminate tomatoes, Swiss chard.

First, mark out the hills, usually on four-foot (120-centimeter) centers. Dig up an 18-inch-diameter (45-centimeter) circle as deeply as possible at each proposed hill center. A person could dig a lot of those spots in one day. The sod in those hills won't be dead yet, but you'll have damaged it.

Next, go find some fertilizer. When the English first came to Massachusetts Bay, the locals there taught them to grow corn by digging a hill and putting a fish in it. So spend a lot of the next week going fishing (do it when you're tired out from digging hills). Collect road kills. Have the kids gather cow pats. Or your own pats. Be creative.

What you'll have to use for fertilizer may need to decompose a bit before the seedlings will be comfortable being too close to it, so put it deep in the center of the hill and sow the seeds around the outside of that hill. The fertilizer will also have to be strong stuff; there isn't time to wait for a lot of low-power vegetation to rot. At the same time you're incorporating fertilizer, you'll be giving those hills another digging. This time remove any surviving clumps of sod, shake off the soil, and toss the roots aside to shrivel in the sun.

In less than ten days you will have started a garden consisting of fertilized hills surrounded by un-dug (as yet) ground. Now sow seeds; put in plants. While you wait for the seeds to germinate, there are four main gardening tasks: kill the grass still growing between the hills by spading the area up, make more hills for sowing autumn-harvested crops, collect more fertilizer, and dig in that fertilizer in ever-widening circles around the hills. Keep digging in fertilizer in ever-expanding circles, keeping ahead of the growing plants so you don't disturb their expanding roots. By the time the season is half over, you'll have dug the entire plot and fertilized most of it.

Except for potatoes. There is a better way to grow survivalist spuds. Dig rows about 12 inches (30 centimeters) wide with the row centers three to four feet (90 to 120 centimeters) apart. The between-row spacing depends on how much rainfall you're anticipating (spacing of spuds will be discussed at length in Chapter 9). Fertilize the rows as best you can and then plant the seed potatoes down the rows. Before the vines emerge, turn over the sod beside those rows extending out about another 6 to 12 inches (15 to 30 centimeters) on each side. If possible, fertilize that soil too, so the spuds will be growing in a loose fertile strip 18 to 24 inches (45 to 60 centimeters) wide. With the rest of the between-row space, dig or chop (with hoe or mattock) enough to kill the grass. As the vines grow, you'll be pulling soil and weeds up against them using an ordinary hoe, gradually hilling up the growing vines. You'll likely end up with a surprisingly large crop, and all the ground in the potato bed will have been cleaned of grass and weeds, ready to use for something else next spring.

Once things are going, there'll be time to start a compost pile and to think about growing more-demanding veggies the next year.

The bow rake

I don't think I could start small-sized, delicately sprouting seeds directly in their growing beds unless I had a bow rake. (I could do it with a "T" rake, also, but for some reason the "T" doesn't seem to balance as well or work as effectively, so if you've any choice in the matter, get yourself the bow pattern.)

Raking levels the surface and creates seedbeds out of already loosened soil, and that is gentle work. The rake handle should be long and slender because the less the rake weighs, the slower your arms will tire. You want the head of the rake to be at least a foot (30 centimeters) wide, otherwise you'll be pushing it back and forth too often to accomplish the same task. Don't get a child's rake with a narrow head and a short handle. At the other extreme, I suspect if a rake were over 16 inches (40 centimeters) wide, it would be too hard to control.

There is a Zen to raking that is not so different from what athletes refer to as "being in the zone." Raking is a meditative action you'll get better at over many years. Raking beautifully requires the same sort of absolute concentration it takes to hit a golf ball, tennis ball, or baseball or to give a good haircut. Your complete attention has to be where the teeth are combing through the soil. Your upper body has to be balanced, your stance as perfect as a martial artist's. That's because the weight that the rake's head is putting on the bed has to be precisely controlled so the teeth comb out the lumps. A rake with teeth longer than two inches (five centimeters) will endlessly be lifting up new lumps to remove. There is nothing wrong with having lumps in your bed if they are not located in the top inch.

Making a seedbed

When a dry seed is placed in moist earth, it swells and begins sprouting after a few days. Any seeds that take more than three to four days to get going are either in soil that is far too cold for them, are in soil that is far too dry to pass the ready-to-till test, or are a slow-germinating species, usually one with a hard seed coat that resists moisture penetration.

By "sprouting" I do not mean that you'll see the leaf emerging at three to four days. In each seed resides an embryo, a miniature, living, breathing plant that is alive, resting in a state of dormancy (in other words, sleeping). When it is moistened, the embryo starts producing complex chemicals called enzymes that convert the dried food surrounding it into water-soluble sugars and amino acids. This is just like digestion in your stomach. The embryo begins to assimilate this nutrition and grows, forming a complete plant with it.

Because drying out is one of the biggest dangers the new plant faces, the first thing that a clever embryo does is to send down a root. With a root system working, there is less chance that the seed will die from lack of moisture before it can emerge. Once the root is extending downward, there is no longer any reason to water the seedbed and a lot of reasons not to.

A sprouting seedling needs a lot more oxygen than it did in dormancy; the embryo will die if there is no oxygen in the soil, which is what happens if the soil is waterlogged for more than a few hours. And like all other biological processes, the speed at which the embryo grows and develops is determined by temperature. If the soil is too cold, the seedling, slowed by the chill, will be attacked by fungi and other soil-dwelling diseases that thrive in cold conditions, and it will die. Watering lowers the soil temperature.

Only after the root is descending does the embryo begin to form what is called the "shoot," the part that emerges into the light. Rarely is this shoot capable of exerting much mechanical force. It does have quite an ability to wriggle its way between larger soil particles, but it can't push aside a heavy lump, much less break through concrete.

Two similar soil phenomena, "crust formation" and "puddling," commonly prevent shoots from reaching the light. Puddling happens when it rains or the gardener waters after the seed is planted, and the soil particles, previously light and fluffy, slump into a solid, nearly airless mass. Puddling usually happens only to soils made up of the finest sorts of silt, or of clay that also lacks organic matter.

Crust formation happens on soils that contain some clay, and since most soils contain at least some clay, crusts can form on almost any land. A crust develops in exactly the same way a cement finisher puts a smooth surface on a concrete slab. Concrete is a mixture of smallish gravel, coarse sand, and cement. While it is still plastic (liquid) the finisher rubs the surface with a trowel or board. This rubbing causes the gravel to sink into the mixture, while the finest sand particles and cement rise to the surface. A thin skin of nearly pure cement blended into the finest sand is formed on the surface. When this dries it feels perfectly smooth.

Soil particles are smaller than the aggregates making up concrete, but they blend much as sand, gravel, and cement are mixed in concrete. Think of the sand in the soil as being similar to the gravel in concrete; the silt in the soil like the coarse sand; and the clay in the soil like the cement in the concrete. Pounding rain (or sprinkler) droplets hitting the soil act exactly like the cement finisher's trowel, separating the clay/silt from the sand. The sand settles a fraction of an inch, leaving clay and silt on top that dries into a tough impenetrable skin.

It is essential that the gardener do something to prevent puddling and crusting from happening. Fortunately this is not difficult to accomplish — simply increase the amount of decomposed organic matter, called humus, in the soil. When enough humus is mixed into the equation, the particles of sand and silt and clay become more firmly cemented into stable, irregularly shaped chunks or "crumbs" of larger size, sometimes as large as grains of uncooked rice. The crumbs don't easily separate into sand and silt and clay, so the soil won't crust over or slump after a rain or irrigation. They also create more air spaces in the soil, greatly improving root development and, thus, plant growth. Soil with an obvious crumb structure is said to have good tilth.

The trouble is that to rapidly change the entire top foot of a poor soil into one that has good tilth can mean blending in as much as two to three inches (five to eight centimeters) of compost or well-rotted manure. That is a heap of compost!

To cover a 100-square-foot (10-square-meter) growing bed three inches (eight centimeters) deep takes one cubic yard (0.75 cubic meter) of material. A long-bed half-ton pickup truck holds one cubic yard — even more if you build up the sides with retaining walls and have strong springs and tires. It could take 18 pickup-truck loads to convert an average 2,000-square-foot (200-square-meter) garden into soil of good tilth one foot (30 centimeters) deep. ■

Fortunately it is not necessary to turn the entire top foot of a garden into the highest-quality topsoil to get a good growing result. Instead it is only necessary to focus on the "seedbed." Most small garden seeds, the ones that have the hardest time germinating outdoors, are placed about half an inch (1.25 centimeters) deep. A few of the larger small seeds, such as beet, radish, or spinach, are sown about three quarters of an inch (2 centimeters) below the surface. So what I propose is that you make only the top inch (2.5 centimeters) of your growing beds into friable soil with excellent tilth. Accomplishing that takes a layer of manure or compost only a quarter inch (six millimeters) thick. Most gardeners find it hard to spread that little. One pickup-truck load would cover fourteen 100-square-foot growing beds a quarter inch deep. Finding and transporting that amount is manageable, even if it must be done with a wheelbarrow.

Once you have your bed covered with a quarter inch of well-rotted manure or compost, use the rake to accomplish four things at once: (1) level the surface

Figure 3.4: *The clods raked down out of a raised bed come along with a bit of compost. These are mixed together as the gardener's feet crush the clods over the coming months. When the bed is reformed for the next crop, this improved soil is tossed atop the bed and will be unlikely to form clods again.*

precisely; (2) uniformly blend the manure/compost into the top inch, using the action of the rake's teeth passing through the soil repeatedly as (3) the rake also breaks up some of the less solidly cemented clods and (4) combs out most of the lumps in the surface layer. The rake will pull these resistant lumps to the edge of the bed, where they roll down the side and harmlessly await the gardener's feet to crush them.

Restoring a raised bed for planting again

After a raised bed or a raised wide row has grown a crop, you will need to rebuild, fertilize, and fit it to grow another. This is the easiest procedure:

- Remove remnants of vegetation to the compost area.
- If the bed or wide row is going to grow medium- or high-demand crops, fertilize it. Fairly uniformly spread four to six quarts (four to six liters) of COF per 100 square feet (10 square meters) of bed or 50 lineal feet (15 meters) of wide row. Otherwise cover the area about an eighth of an inch

Measuring soil amendments

- A 100-square-foot (10-square meter) raised bed is 4 x 25 feet (120 x 760 centimeters) or 40 inches x 30 feet (100 x 900 centimeters).
- 50 lineal feet (15 meters) of wide row (two shovel blades wide) is 100 square feet.
- It takes one cubic foot to cover 100 square feet (20 liters per 10 square meters) an eighth of an inch (three millimeters) deep.
- To cover 100 square feet a quarter of an inch (six millimeters) deep takes two cubic feet.
- To cover 100 square feet half an inch (12 millimeters) deep takes 4¼ cubic feet.
- Five gallons (20 liters) equals one cubic foot. The large plastic pail that contains bulk foodstuffs usually holds five gallons.
- A cubic yard is 27 cubic feet (0.765 cubic meters). A long-bed pickup truck can carry the weight of a cubic yard of manure or compost with no problems.
- Fifty pounds (25 kilograms) of lime per 1,000 square feet (100 square meters) is about a quart-jar full (one liter) per 100 square feet. ■

(three millimeters) deep with strong chicken manure compost (containing over 3 percent nitrogen).

- With the shovel, scoop up about an inch (2.5 centimeters) of soil from the paths around the bed or between wide rows. Toss that soil atop the bed. If the paths have gotten a bit weedy, this action will clean them. A sharp shovel wielded by a pair of strong arms can cut off weeds as the soil is lifted. Otherwise, scrape the path with a hoe first and then lift the soil. The hoeing will also loosen the soil in especially compacted paths, making the rest of the task easier.
- Roughly turn the bed over to the depth of the shovel's blade. This mixing need not be thorough. It is okay to stand on the undug part of the bed while doing this. At all other times keep your feet off the bed to avoid compacting the growing area.
- If much soil ends up back in the paths from this digging, scoop it up and toss it onto the bed again.
- With a rake, roughly level the bed.

Figure 3.5: *Restoring a raised bed for planting.*

Figure 3.6: *Screening compost to separate the fine from the not fully decomposed material. A large sieve like this is one of the most useful homemade tools a gardener can make.*

- If you are going to sow small-sized seed, cover the bed with a quarter inch (six millimeters) of well-decomposed manure or fine compost. Do not use clumpy half-rotted compost; the texture must be fine. Some gardeners have a screen to sieve out fine compost for this purpose. If you're going to sow large seed, using rough compost will serve just as well.
- Now take a rake and perfectly level the bed, pull the clods out of the surface inch and down into the paths, and make a seedbed ready for sowing. If you aren't going to sow fine seed, this last step can be done quite carelessly.

It takes me about one pleasant hour to accomplish this task on a bed of 100 square feet (10 square meters) — spreading potent manure, compost and/or COF, spading the bed, scraping the paths, sieving compost, hauling it to the bed and spreading it, and raking out a seedbed.

The hoe

I have a tool collection in my tool shed, with a variety of hoes that are extraordinarily efficient for doing one particular task. But only one of my hoes is good for everything — naturally, this one is the most common design. Most people call it a "garden hoe." The rectangular blade is attached to a swansneck, so called because when held in a vise, and with the application of a fair amount of force, this rod of mild steel can be bent to adjust the angle of the blade so it matches the height of the user. The swansneck should be bent so the blade is parallel to the soil when the handle is held in a comfortable position with the user standing upright. At this angle a sharp blade can rapidly and easily cut off weeds just below the soil's surface, while at the same time loosening up the surface and breaking any crust that may have formed. This loosening action is called "cultivation." If the angle is wrong, so that the hoe wants to dig itself into the earth when pulled toward you, then you can't efficiently work the tool because you will have to hold the handle too far from your body in order to correct the angle of the blade, which makes you tired. If the blade wants to lift itself out of the earth when pulled toward you, then you will have to bend over to adjust the angle of attack, which also makes you tired. Once the swansneck has been bent to suit the user, it never has to be adjusted again.

Sharpening the hoe. File the outside of the blade so it makes a sharp, uniform, chisel-like bevel of about 20 degrees. To sharpen it the first time, put the hoe on the ground or on a porch so the blade is at the edge of the porch. With one foot placed close to the socket, press down with enough weight to hold the blade still while you file the outside of the blade, holding the file with both hands and bending your body over it. If this is too difficult, have someone else stand on the handle or else put the blade into a bench vise.

After a proper bevel has been created the first time, you may be able to resharpen the hoe by filing one-handed. With the bottom of the handle wedged into the ground and restrained by the edge of your shoe, lean the blade against a stout post at least four feet (120 centimeters) tall and with one hand press the back corner of the blade firmly against the support. File. It shouldn't take a sharp file more than ten or so strokes to restore the edge.

With a sharp garden hoe, you can perform the following tasks:

Figure 3.7: *Hoeing technique for weeding soft ground.*

- **Weed soft ground.** Pull the blade toward you, half an inch (1.25 centimeters) below the surface. This cuts off weeds as though with a knife. Most weeds will die within hours of being cut. The odd one that does not die because it was not completely separated from its root system will almost surely die when you cut it again a week later. Weeds that come back from tubers or bulblets may require cutting six or eight times over six or eight weeks. But there is no weed in Creation that won't die if its leaves are repetitively removed before they can make much food — even horseradish or comfrey eventually succumbs.
- **Weed compacted soil** — including paths. When weeding compacted earth, you hold the hoe entirely differently. Keep the handle slightly below your waist. Place one hand about halfway down the handle and use this hand to press the blade down while the other hand pulls the sharp blade toward you. The blade scrapes across the earth; because it is sharp, it cuts off small weeds. If your hoe is truly sharp, you'll be surprised how quickly this can go. This action tends to dull a hoe faster than any other use. When I have a lot of paths to scrape, I take my file to the garden.

- **Chop out resistant clumps of grass and large weeds.** When doing this, use the corner of the blade. Swing the hoe like a mattock or axe; the corner penetrates the hard earth, chopping off tough stems. This only works well when the corner is also more or less a proper right angle and the working edge is kept sharp. Even then, this task is exhausting hard work. If the garden is weeded once a week during the growing season, there will never be large, resistant weeds to chop out. All of them will be killed rapidly and with little effort when they are small and tender. I can weed 2,000 square feet (200 square meters) of growing beds and the paths between them in about one concentrated hour. I do this about once a week during

Figure 3.8: *Hoeing technique for weeding compacted ground or scraping paths.*

the height of the growing season, and less often as summer wanes because everything, including the weeds, grows more slowly then. Spring is the most critical time to hoe weeds because that is when they are growing rapidly and because unsettled weather often keeps the gardener out of action while weeds grow to the size that demands exhausting chopping rather than easy scraping.

- **Perform other tasks.** The ordinary garden hoe can do a few other things. A corner of the blade can be used to make furrows by pulling it through soft soil. With a bit of practice, the depth of that furrow — a factor critical to successful germination — can be held quite uniform. Then, with seeds in the furrow, you can use the flat of the blade, front or back, to shove some soil back into the furrow, either on the push or on the pull stroke. (In Chapter 5 I will suggest a better way to use the hoe for making shallow furrows.) This ordinary hoe can also pull earth toward the gardener, which is useful for hilling up potatoes. However, if a large plot of spuds is to be grown, using a larger, heavier version of this hoe will save a lot of time.

Hoe quality

I once foolishly purchased a cheap hoe. I still owned that disgusting thing when I went into the seed business, and because I had no extra money during my first few years in business, I used it to weed the trials ground. The steel was so poor that by the time I reached the end of each 100-foot (30-meter) row it was dull. It takes at least three times the effort to weed with a slightly dull hoe; the effort it takes to weed with a really dull one can't be described in polite English. So I took to carrying a file in my back pocket, and when I completed each row I would lean the hoe against a fence post and resharpen it. After two years there wasn't much blade left to sharpen.

Then I got a much better-quality implement. Same general appearance; totally different tool. The quality hoe gets sharpened each spring and stays sharp for about half the summer unless I am scraping a lot of paths with it.

Hoes have a fairly reliable quality indicator you can see before you purchase. The tool is intended to occasionally chop resistant weeds. This action exerts a big shock on the point where the swansneck attaches to the handle. There are two ways to attach a swansneck. The cheap way is to simply push the rod into a hole in the handle and then reinforce that basically weak joint with a sleeve or collar of thin metal (a ferrule) that fits tightly around the wood so that (the manufacturer hopes) the handle won't split when some real stress is applied. But even if the first powerful shocks don't split the end, they do gradually enlarge the hole so the swansneck loosens up in its socket and the hoe head starts wobbling or rotating. The owner then inserts all sorts of wedges or carries out make-do improvisations to tighten up the tool. Eventually he or she has to get a new handle.

There is a better way. The end of the swansneck can be forged (or welded) into a strong, slightly tapered socket that is slid over the end of the handle and affixed with one or two screws or by a bolt that goes through the whole thing. Any manufacturer with the integrity to make a forged metal socket to attach its hoe to the handle is almost certainly going to use quality metal for the blade.

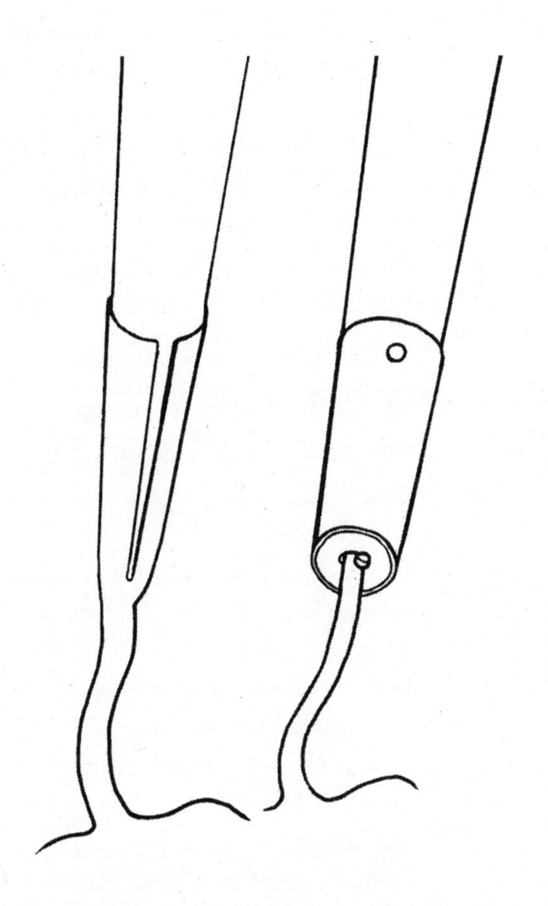

Figure 3.9: *Notice that the swansneck on the right has loosened and a nail has been driven into the hole to function like a wedge in the handle of an axe.*

Push-pull hoes

There is one special hoe design well worth having. The push-pull hoe works by gliding back and forth through soft soil, cutting weeds off just below the surface on both the push and the pull stroke. There are two common designs: the stirrup hoe (or "hula hoe") and the propeller hoe (or "glide and groom"). These hoes are available through mail-order seed companies or other mail-order garden gear suppliers. I've owned both sorts with equal satisfaction.

If run over the beds once a week, this kind of hoe will keep even a large garden beautifully manicured in little time and with little effort. The design protects your plants from being accidentally cut off while allowing the cutting edge to pass close to the vegetables. This is especially true of the stirrup design. However, the stirrup design attaches to the handle about four inches (10 centimeters) above the top of the cutting edge, making it harder to pass this hoe under overhanging leaves. The propeller type will slide better between rows whose leaves are getting close to forming a canopy. Every design of every tool involves compromise.

Weeding and plant spacing

Eliminating weeds with a sharp hoe is much quicker and easier than yanking them with dull fingers. You're standing up instead of bending over. You're eliminating little ones by the dozen per motion instead of one by one. When you routinely cultivate the entire bed, you're killing many of them before they've even grown large enough to see. The catch is that to do it this way, the rows must be far enough apart for the hoe to pass graciously between them. That means a minimum 12 inches (30 centimeters) between-row spacing.

When I am growing small plants like radishes, finger-sized carrots, or salad turnips intended to be pulled hardly larger than radishes, then the rows are 12 inches apart with the plants thinned to one to three inches (2.5 to 8 centimeters) apart in the row. The rows are made across a wide raised bed, so each row is 3½ to 4½ feet (100 to 140 centimeters) long. For a single sowing I might start one or more such rows. Larger plants like beets, storage carrots, and parsnips grow better in rows separated by 18 inches (45 centimeters). With this spacing I can keep the weeds conveniently hoed until the leaves of one row begin to touch and interpenetrate the leaves of the next row. Then it is no longer possible to slide any sort of hoe between the rows. Once a leaf canopy forms, it strongly shades the soil below. Even if a weed should sprout, the canopy prevents it from growing much. Only the odd weed will show itself above a crop canopy, and this one you can yank out by hand.

If I were without irrigation, dependent on rainfall, and needing to be able to get the garden through rainless periods lasting several weeks, I'd increase the between-row spacing and thin the plants a little further apart in the row. In that event I can continue hoeing through the entire growing season. And with a push-pull hoe, especially of the propeller design, I can usually get right up to the vegetables' stems. Thus there need be next to no hand-weeding in a garden.

Chapter 6, on watering, will give you plant-spacing charts for all these circumstances.

Miscellaneous tools

There are myriad garden tools and gizmos offered for sale, but the other items you really can't do without are a wheelbarrow, a sprayer, knives, and buckets.

Wheelbarrow

Garden magazine advertisements tell you how labor-saving it is to have an expensive two-wheeled garden cart instead of an "old-fashioned" wheelbarrow. Unless you are practicing no-dig gardening, there is no need to be hauling dozens of pickup-truck loads of organic matter around the veggie garden. A large wheelbarrow holds enough compost or aged manure to adequately restore a bed of 100 square feet (10 square meters) — plenty of hauling capacity. The wheelbarrow also maneuvers neatly down narrow paths between beds that a two-wheeler couldn't manage. And a wheelbarrow costs a lot less money.

But do get a good one, with strong wooden handles and a broad pneumatic tire (with inner tube) on a steel rim, turning on ball bearings. No plastic! Get the largest barrow of the best quality you can barely afford. To find it you'll probably have to visit a farm supplier or commercial hardware, not the local branch of a national discount chain. Make sure that the handles are high enough for your body. When you lift the handles and hold the "barrel" so it barely tips forward, your arms should be only slightly bent at the elbows and your back nearly straight. There should be room for you to stand between the handles and walk forward without risking your knees. A proper fit to your body is essential! Otherwise you can't control the barrow when it is heavily loaded, and if you have to bend over to push the barrow, you'll make your back sore. There can be a lot of difference from one make to another; try several and find one that fits your body.

Don't scrimp on this purchase. Once you have a wheelbarrow, you're probably set for the next 20 years if you keep it painted so it won't rust away.

One other thing: Builders' supply companies sell specialist wheelbarrows designed to carry concrete. These have a rather deep, rounded chamber and thick-walled tubular metal handles instead of wood, are super-strong, and are too heavy for gardening. This design will make you unnecessarily tired if you use it much. The style you want will be capacious but lightweight. Most of what you're going to carry is not dense like concrete or gravel, and you'll want to be able to carry as much volume as possible.

Sprayer

There will be times that you'll want to spray — liquid fertilizer on leaves; compost tea to fight disease; *Bacillus thuringiensis* (Bt, an organically approved pesticide) against corn earworms or cabbageworms; assorted home remedies like soapy water, rhubarb leaf juice; etc.

The sprayer you buy should not be a cheap one, should hold up to five or six quarts (five or six liters), and should be the sort you pump up and then carry around by the pump handle. As an indication of quality, take a look at the nozzle. The best sprayers use interchangeable, inexpensive nozzle inserts, small in size and usually made of brass, so that different volumes and patterns of spray can be selected. The sprayer that has a built-in, adjustable cone nozzle made of plastic is usually a cheapie — but not always. Ask if the seller stocks any spare parts, like pump pads and o-rings. Also look at the pump rod. If it's made of plastic, it may become brittle after a few years and snap. It's better if it's made of metal.

But most sprayers these days, even the better ones, are made of plastic. Protected from the sun, a plastic sprayer might last a decade, but one left in the sun will survive two or three years, tops. It's sad. We could be building gear to last. Maybe when we seriously run short of oil, plastics will start costing more than metal. And we'll again have equipment that'll last a lifetime.

One last tip about sprayers: Make sure the stuff you put in them is free of particles and that no grit is lurking around the pump's seal before you open the tank. Rinse it off/wipe it off before unscrewing the pump to fill the spray tank. If particles get into the tank, the dratted thing will likely get plugged up.

And finally, whatever you spray won't do much good if it doesn't stay on the leaves. Many plants have waxy leaves that make water bead up and run off. To keep your spray on the leaf until it dries or penetrates, always use a surfactant, which is a fancy name for a water softener. The cheapest effective one is a quarter teaspoonful (1.5 milliliter) of ordinary mild liquid dishwashing soap per quart (liter) of spray. I've never had this quantity burn or otherwise damage leaves. Don't mix a strongly antiseptic detergent with Bt.

Small kitchen knife and medium-coarse stone

I keep two identical thin-bladed, sharp, pointed, kitchen paring knives stuck into the top of a short wooden stake in the garden. Next to the stake is an inexpensive large, double-sided (medium/ coarse) sharpening stone. The knives have bright red plastic (highly visible) handles because I'm always trying to remember where I last put one down. The stone stays in the garden because knife steel isn't made for running through the earth while thinning tiny seedlings or weeding close to nearly emerged rows. I resharpen a knife after every five or ten minutes of use. The stone will, in an emergency, also hone a dull hoe.

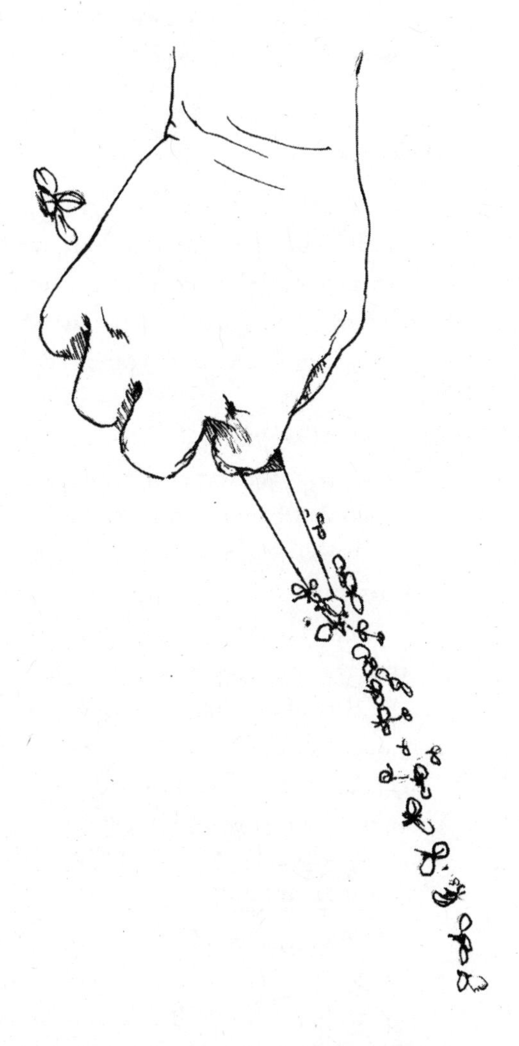

Figure 3.10: *Using a paring knife for thinning or weeding.*

A thin pointy knife will slip between crowded seedlings, allowing me to cut some off, permitting others to grow. If I tried to do this sort of close work with my fingers, it would take three times as long. The knife will snick off the odd weed just below the surface should I notice it while doing other things. I also use the knife for harvesting.

I could make a garden without a hoe. I wouldn't want to make one without a rake. I couldn't make a good one without a shovel and definitely not without a small sharp knife.

Buckets

Buckets do more than carry things; I use them for fertigation. I'll have much to say about fertigation in Chapter 6 on watering. A few cheap plastic ten-quart (ten-liter) laundry buckets or, better, some recycled five-gallon (20-liter) ones that once contained institutional kitchen bulk goods are essential around the garden. It seems one can never have too many.

Care of tools

Once tools were expensive, valued, and cared for. Then we created a consumer society in which nearly everything is designed to decompose so another will be purchased. Maybe once resources become more scarce we'll go back to the old ways. But in the meantime, people have forgotten how to care for valuable tools.

Two things go wrong with every shovel, rake, and hoe. The metal bits rust; the wooden bits dry out, crack, and split.

Rust. In the old days, there was usually a wooden box on the floor of the garden shed filled with coarse sand. The sand was saturated with used crankcase oil. When the garden tools were put away for the day, their working ends were first wirebrushed clean and then stuck into that sand so they came out lightly coated with oil. These days people sometimes have a convenient spray can of lightweight oil that serves nearly as well for this purpose.

Handles, wood. I have a dusty pint (half liter) jar half full of coconut oil in the shed. At the end of the season, when everything is being put away for the winter, I bring that jar into the house to warm it so it becomes liquid. Then I thickly coat my wooden tool handles with this oil, making sure plenty gets into the sockets where blades attach to the handles, and hang the oily tools up on the wall, suspended between pairs of large nails. It takes a while for all that oil to sink in. This once-a-year treatment seems sufficient; the handles don't roughen and/or split and/or shrink unduly. No other common vegetable oil works as well as coconut oil, which I buy at a health food store. I've seen some fancy preparations made for protecting tool handles, but basically they're a more-or-less-equal mixture of coconut or linseed oil and beeswax.

Handles, composite. One of my shovels has a composite handle, which is lighter in weight and stronger than wood. However, composites don't much like being exposed to ultraviolet light. Keeping it out of the sun when it's not in use does help the handle outlast the blade.

CHAPTER 4

Garden centers

On spring weekends, garden centers are so busy that people stand in long checkout lines holding armloads of expensive seedlings and a few seed packets. Usually these buyers are making several major mistakes at once. What I'm about to tell you about those errors won't endear me to garden center owners and makes it highly unlikely that my book will ever appear for sale in those places. Oh well ... personal honor is infinitely more valuable than bigger royalties.

Let's take a stroll through an imaginary garden center at the peak of spring planting season and discuss what we find.

The mind is strange. I direct my magic keyboard to an imaginary North American garden center in spring, and instead I am taken to scenes from my childhood in Michigan. Winter is ending. The snow melts. The earth thaws. Some weeks later, on a sunny afternoon, I am walking along a street with my jacket unzipped, and for the first time there appears an intoxicating odor — spring. The soil has warmed up enough that I can smell the bloom of soil fungi. The nose of this spring wine is rich, fruity. When I was young and this aroma hit, some instinctual joy would grab me and I would go running down the street, block after block, fast as I could.

You'd expect on such an especially wondrous day that sedate older people would be out digging their gardens. But no, they may admire the tulips by the front door, but their backyard veggie gardens are still growing weeds instead of spinach, radishes, and peas. Weeks pass. Finally the much-spoken-of day of the last usual frost arrives. Now people mob the garden center, buy an instant garden, rush home, start up the rototiller or start digging up the backyard, and "put in the garden."

Putting-in is usually done over one weekend, sometimes in one afternoon. Typically tomato, cabbage, broccoli, onion, lettuce, zucchini, cucumber, pumpkin, and winter squash seedlings are taken home and set out, all in a few hours. Some people even buy sweet corn as seedlings for transplanting. A few seeds are sown: carrots, beets. Oh, what joy and hope my imaginary people are feeling! But I don't know whether to smile, laugh, or cry. I like it. I hate it. What a waste. What fun.

My emotions conflict because I know what happens next. Some of the seedlings fail to survive, so more are purchased and replanted the next weekend. Sometimes this repurchase/replanting is repeated yet again. A great deal of money is invested in this hobby garden. The folly is made worse when people buy seedlings for types of vegetables that should be started directly from seeds, but many gardeners believe it is impossible to make seeds come up.

And then what happens? All six cabbage seedlings from one little tray survive. Two months later, six heads harden on the same day, and one week after heading up, five of them, not yet cut, begin to split open. But it is too hot in high summer to make good sauerkraut, which isn't wanted anyway. Coleslaw is what is wanted during the months of August and September. So, to make the best of it, all the neighbors or relatives get a splitting cabbage.

Broccoli? Same thing happens. Six big flowers form in the same week. This bounty gets frozen, but no one in the family likes eating frozen broccoli much. Even if they did, the variety in the tray doesn't freeze well, so it mostly goes uneaten. After the large central flowers are cut, only a few small woody side shoots form because the variety selected was bred to be grown for the supermarket trade, to quickly produce one central flower and be done with it. A different variety could have made large succulent secondary flowers for several months after harvest of the main head. And there are varieties of cabbage bred to stand for many weeks after heading up before they split.

Luckily there will be some big successes; there are so many tomatoes to pick that the pantry is packed solid with sauce and green tomato pickles.

Come the end of the season, the pantry work shelf is covered with ripening tomatoes that last into autumn. No wonder everyone loves to grow tomatoes.

And next? Next comes frost. The first light frost wipes out almost everything except the Swiss chard. And the gardeners add up what was spent and the worth of what was actually eaten or put by and wonder why they bothered. But next spring the same compulsion that makes a young man run for joy has that same person back in the garden center, doing it all over again.

The next few chapters of this book are mainly about avoiding mistakes in judgment and expanding your imagination, because the garden can be more than fun and a joy and a promoter of health and a producer of food more nutritious than could be purchased for any price — the vegetable garden can also be an astonishingly sensible economic venture. Even where the winter is hard, the garden can supply the kitchen for several more months than most people expect. But for all those good things to happen, the garden has to be given the same degree of attention that other enthusiasts give to selecting the right fishing lure, modifying their automobile, or refining their golf swing.

Can you do that? Can you give food gardening its due attention in this era of the end of oil, this time of a globalized labor pool, this season of resource hostilities? Can you afford not to? If you agree to continue along with me in this book, the first thing I'm going to do is wean you off the garden center.

Transplants: Buyer beware

Whenever I inspect vegetable transplants, I am deeply suspicious. Has the seedling been properly hardened off? Is the seedling pot-bound or the opposite, insufficiently rooted? Is the variety marked on the label useful in the home garden? Is the seedling actually the variety marked on the label? Should this species or variety even be transplanted in the first place? Let's take up these concerns, back to front.

Should it be transplanted in the first place? Some species are extraordinarily difficult to transplant, such as beets, carrots, and chard (silverbeet). But they can look so jolly in the seedling tray! Transplanted carrots may survive to grow a top, but usually fail to make useful roots. Cucurbits (melons, pumpkins, zucchini, cucumber, squash) don't transplant well, and if they don't die from the attempt, their growth is usually set back so severely that seeds of the same variety sown on the same day as transplants are put out will yield just as soon or sooner. So why waste so much money? Lettuce, too, generally does better started from seed; the transplanting process hugely shocks its root system. This is also true of transplanting corn.

Is the seedling mislabeled? Unless it is a red cabbage seedling labeled as a green cabbage, you can rarely tell if the label is incorrect. But please consider this: the wise buyer always imagines the ethical temptations the seller might have and then is not surprised if a seller falls prey to moral weakness. For example, those pretty plastic labels on seedlings cost as much as or more than the seed did, especially if they carry a picture of the fruit or plant on them. The greenhouse operator may have invested in a big box of labels for some variety whose seed is not available that year. What might be done in that case? Wouldn't it be tempting for the seed vendor to get a similar variety and substitute it? Who could tell? Or suppose the seed merchant supplying that bedding-plant raiser was out of stock on some popular variety. Wouldn't it be tempting to make a substitution, "accidentally" mislabeling a bag? Who could tell?

If such a substitution were truly a nearly identical variety, then almost no one would notice. But what if a cabbage variety were popular because it could stand without splitting for weeks after heading up and also tasted nice? What if it were replaced with a cannery sauerkraut variety that split within days and had a texture and flavor like cardboard? That wouldn't be so nice.

One species that isn't likely to be mislabeled is tomatoes. That's because people look forward to eating a familiar fruit every year. It would be hard to fool a lot of customers with tomatoes. Green bell peppers and eggplants, however, could be another matter altogether.

Is the seedling a home-garden variety in the first place? What if, in the crunch of spring, the garden center owner can't get enough seedlings from the usual supplier and so turns in desperation to a seedling raiser supplying farmers growing for the supermarket produce department or the cannery?

What if the seedling raiser figures that corn is corn is corn and you can't tell anything at two weeks old, so any cheap seed tossed into the trays will be suitable? What if the seedling raiser, flinching at paying several cents for each high-quality broccoli seed, offers a much cheaper variety for which the seeds cost only a few cents per hundred? What if? The result will be a broccoli patch in which about half the plants fail to make a large central head and whose flowers are coarsely, loosely budded and likely of poor flavor.

What if any or all of the above happens? Well you, the soon-to-be-disappointed customer, have invested a lot of work, fertilizer, effort, and hopes.

Is the seedling pot-bound? Ah, at last, something about transplant quality the buyer can see before purchasing. Raising transplants for profit is not easy. The seedlings rapidly grow from being too young to ethically sell to being almost too old to ethically sell in under ten days.

When they are too young, their root system has not yet filled the pot; the soil ball around the roots will not hold together when it is transplanted. Should the soil crumble away during transplanting, so much damage may be done to the delicate root hairs that even if the seedling survives, it will not grow well for a week to ten days. That's one reason (of several) that directly seeded plants will often outgrow transplants.

When the seedling has been in its pot too long, the roots will overfill the pot. The pot-bound plant may, if fed liquid fertilizer and watered several times a day, continue growing and look okay above ground, but its roots wrap themselves around and around the inside of the pot. When the seedling is transplanted, that constrained root system can't support the top in hot weather unless it's watered twice a day. This seedling wilts easily until it starts putting roots into new soil. The leaves of pot-bound seedlings will hardly grow for a week to ten days after transplanting.

Why would you spend a lot of money buying seedlings to jump-start the growing season if they aren't going to grow for a week to ten days after you set them out?

A seedling perfectly ready for transplanting will have extended the tips of many roots right out to the extremities of its container, so its soil ball will hold together during transplanting. This seedling can grow from the first day it is put out.

So take a look! Place the stem of the plant between your second and third fingers, hold your palm against the soil, turn the pot upside down so your hand supports the soil, tap gently on the side of the pot, then lift the pot, exposing the soil, and see if you can see any root tips. If none are visible, the soil will be crumbling into your cupped hand. In that event, carefully repack the potting mix and set the plant back down. Let someone else buy it. If you do buy it, be prepared to keep it alive in that pot for about a week, waiting for the root system to develop enough to withstand the mechanical stresses of transplanting.

How did that unbalanced seedling get that way? Isn't the root system's development supposed to match the growth of the top? Yes, it is. But the owner of the greenhouse raising those seedlings was faced with ethical temptations. Profit in that business is determined by how long the seedling will occupy greenhouse space before being taken to the sales bench. The raiser can accelerate top growth — which is what the buyer sees, which is what leads to the sale — by making the greenhouse warmer at night and using a fertilizer balance that pushes top growth at the expense of root development. Or the nursery can sell a variety bred to look good at four weeks old — profitable for the seedling raiser, but not necessarily good for the gardener.

If the seedling is pot-bound, you'll see it right away. Don't buy it. How did it get that way? The seller kept it looking good for a week or more after it should have been sold by feeding it fertilized water and watering it several times a day.

Soft seedlings? When plants are grown at high temperatures, particularly at night, they grow lushly — leaves and stems get much larger — but a goodly part of this size is nothing but water stored in swollen, weak-walled cells. When plants don't experience wind, they are mechanically weak. The gentle battering caused by light breezes actually exercises the plant and causes it to reinforce its connective and structural tissues. But if a seedling diverts its energies into making strong structural tissues, it won't be as large or lush looking. Many seedling growers crank up the heat (there is no wind inside a hothouse) and then, after four weeks, move these attractive-looking seedlings directly out to the sales bench.

Then someone buys this pampered plant and puts it in the soil. On its first night outside it gets chilled, something the seedling never experienced before. This

Quote from Stokes' 2005 catalog listing two similar varieties of Chinese cabbage:

> **76 MICHILI 78 days.** The standard open pollinated tall cylindrical strain used by bedding plant growers ... We recommend hybrids for commercial crops. 1 lb: $7.60
>
> **76G Monument 80 days (F1 hybrid).** Taller, much later than ... slower to bolt ... l lb, about $70.00 (price approximate because quality hybrids are sold by the 1,000 seeds).

Item 76 will produce a high percentage of off-types and non-heading plants. Item 76G will produce a row of heads as similar as peas in a pod. Stokes is telling us that typical bedding-plant growers are selecting their seeds based on cost, not on the results obtained from growing them to harvest. ■

chilling, nowhere near frost but still nippy, is a severe shock that sets the seedling back so hard it can't grow for a week. The next day is cool and overcast, and the wind gets a bit gusty. The seedling never experienced wind before, so its leaves and stems get badly bruised, if not torn and shredded. Another shock; another setback. Next day the sun comes out full blast. The seedling never experienced unfiltered sunlight before, having always lived under plastic that filtered out about 15 percent of the energy. Another shock. In that strong light, its underdeveloped (or pot-bound) root system can't bring in enough water and it wilts a bit ... another severe shock. And then some disease or insect that wouldn't normally harm a healthy plant takes it out. Or perhaps the seedling manages to run the gauntlet of hardening off (adjusting to normal conditions) and finally gets to growing right, but it might take two weeks to get past this hardening off.

You will know it was a soft seedling (if it survives) because it'll make much smaller leaves a few weeks after transplanting than it had when you purchased it. A properly hardened seedling that looked smaller and more wiry when it was planted out at the same time will be way ahead two weeks later.

Ethical seedling raisers harden off their seedlings by moving them from their mimimally heated greenhouse to what is called a cold house when they are about three or four weeks old, then growing them on in this more natural environment for a few weeks until they have toughened up. A cold house consists of a plastic roof and enough side walls to somewhat break the worst of the wind if it should get to blowing strongly. The structure is designed to raise nighttime temperature a few degrees but still allow the seedlings to get a bit chilled. The seedlings end up smaller, and the stems and leaf stalks look a bit like an endurance athlete in tip-top condition — no fat and with corded muscles showing.

On the sales bench and given the choice, the average buyer will choose the soft, tender, lush seedling over one that has been hardened off properly. And thus the ethical temptation is doubled because hardening off lessens salability and, because it takes another few weeks, it also costs the raiser more.

What's a gardener to do about all these risks? If you're buying, make sure you can trust the seller. Grow your own seedlings? My answer: yes and no.

Grow your own, yes, but more importantly, be they yours or grown by someone else, don't transplant plants unless absolutely necessary. And in Chapter 5 I'll show you that transplants are rarely necessary.

Growing your own seedlings the easy way

I suggest you focus on growing vegetables that can be directly seeded in spring. Yes, transplanting allows you to harvest some chill-hardy crops a few weeks ahead of those you direct-seed, but this takes a lot of work, and if you adjust your attitude a bit, this effort will seem unnecessary, even a bit foolish. Accept my advice for this year, at least, and you won't need to grow or purchase more than a few dozen seedlings. You won't need greenhouses, hot frames, heat mats, seedling trays, etc. And your directly seeded vegetables will come up handsomely because you are going to follow my advice (in Chapter 5) about where to buy strong seeds and do what I suggest to make them perform. If you want to push the limits, there is no shortage of books that'll help you build greenhouses and hot and cold frames; there are nursery suppliers that'll outfit you with the neatest of seedling-raising gear. But to simply and inexpensively get a lot of good food in every month it is possible to get it, none of these exertions and expenses and stresses are needed.

My advice is to grow seedlings only for those fruit-producing species that benefit from being given every possible frost-free day: i.e., tomatoes, peppers, and eggplants. If you garden in a short-season area, you may want to start a couple of melon or winter squash transplants. Here's how to do it yourself.

Soil for raising seedlings

Do not purchase potting mix. It is often not soil but decomposed bark and other woody wastes with some chemical fertilizer added. Shade-tolerant indoor ornamentals (plants adapted to growing on the floor of a tropical forest) will thrive in this kind of humus-based medium, but vegetables, which are sun-loving plants, grow in soil. Buying a soil mix correctly compounded for raising seedlings, and made fertile with rich compost, is also a waste of your money. Your seedlings may as well start out in and get used to the same soil they are going to grow to maturity in.

You should have no need for sterilized soil. True, by using a growing medium free of bacteria and fungi you can sometimes coax old, weak seeds into life. But you are going to use strong, vigorous seeds that don't need coaxing. In a sterile environment, most of the seedlings that come up will thrive. This is profitable in commercial applications. But you are going to plant several seeds for every plant you'll ultimately grow, so a few losses due to soil diseases won't matter. In fact, a few losses will be to your benefit because the soil pathogens will eliminate the weaker seedlings for you. The ones that thrive in your own garden soil from the beginning will likely continue to thrive to the end.

To make seedling soil, get a five-gallon (20-liter) plastic bucket and half fill it with ordinary garden soil. If it isn't clayey soil, blend in about 1¼ gallons (five liters) of well-rotted manure or well-ripened compost. If your garden is new and you haven't made any compost yet, buy a sack of compost. Beware of using bagged feedlot-steer manure for this purpose; sometimes it contains ordinary salt. If real compost is not available, use sphagnum moss.

If your garden soil is clayey, you'll definitely require sphagnum moss. If you try to lighten clay up with manure or compost, you'll probably make the mix too rich to grow good seedlings. Instead, thoroughly mix into clay an equal volume of well-crumbled moss. One small bale should be enough moss to last you through a great many years of seedling raising. I suggest moss because it contains almost no plant nutrients and doesn't decompose rapidly. It will make an airy mix that'll stay loose for several months.

So far, what we have done is create an airy, loose growing medium. But unless you made it with super-good compost, the mix will lack mineral nutrients. The best way to fortify that mix is to add exactly one cup (250 milliliters) of complete organic fertilizer (see Chapter 2) to each three to four gallons (12 to 16 liters) of seedling mix. If you have no COF, it is essential you boost the medium's calcium content, so add a quarter cup (60 milliliters) of agricultural lime. Measure accurately. Blend it in. You should end up with a five-gallon (20-liter) bucket about three-quarters full of fertile seedling mix, enough to fill two or three dozen substantial pots.

Pots

Improvise. Waxed paper or plastic milk containers with a few drainage holes poked into the bottom make acceptable seedling pots. It's best to use the lower third of the one-quart (one-liter) size. The bottom four inches (ten centimeters) of large-sized waxed paper or Styrofoam beverage cups are good. Or use

Figure 4.1: *A seedling pot made with a strip of newspaper and a rubber band or string.*

leftover plastic pots from earlier years when you were foolish enough to buy seedlings. Whatever you use, the seedling pot should hold a bit more than a half pint (250 milliliters) of soil. For starting the fast-growing cucurbit family, which have delicate roots easily damaged by transplanting and which are best grown indoors for no more than one week, make a pot using a three-inch-wide (8-centimeter) strip of newspaper rolled into a squat cylinder, filled with soil, and held together with a rubber band and/or a piece of string. Once filled, the pot must rest on a cookie sheet or something similar until the seedling's roots have filled the soil, allowing you to pick it up without the earth falling out the bottom. This sort of "pot" is especially useful because the whole thing can be planted without damaging the roots in the slightest. In my own cool-summer climate I use pots like these for starting melons.

Sowing and sprouting

The soil should be at the perfect moisture content for germination before you sow the seeds because you should not add any water until germination is done. This happens to be the same moisture content you'd want soil to be at before you started digging it. Use the ready-to-till test described in Chapter 3. If the soil is too wet, you'll end up with considerably poorer germination.

Fill each pot to the top with loose soil and then press down gently with your palm so there aren't any big air spaces. The soil should end up about a quarter to a half inch (6 to 12 millimeters) below the rim. With the eraser end of a pencil make one small round hole, three eighths of an inch (nine millimeters) deep, in the soil at the center of the pot. Count out four or five seeds and drop them into the bottom of the hole. Press them gently down into the bottom of the hole with that same pencil, but don't deepen it. Flick a bit of loose soil into the hole to cover the seeds; this way there is nothing weighty to oppose the emergence of the shoots. Now slip a small, lightweight, thin, clear plastic bag over that pot. The bag should be large enough that it can be twisted closed, though I always put the extra plastic under the pot and hold it shut with the weight of the pot on top of it. Why do you use a baggie? Because each time you water a sprouting seed, the medium gets too wet and the moisture makes the temperature drop. Neither of these is helpful. Inside a nearly airtight bag there is almost no moisture lost. The soil starts out at the perfect moisture level and stays that way until the seedlings emerge.

Start solanum seeds (tomato, pepper, eggplant) about six weeks before you'll want to transplant. Sow the tomatoes first. When these are up and growing, start pepper and eggplant seeds. When these are up and growing in your sunny window, consider starting the hardiest cucurbits: first the zucchini and squash. A few weeks later, after outdoor conditions have become warm and settled enough to put out the tomatoes, think about starting cucumbers and melons. Cucurbits take less than two weeks from the first sowing of seeds to the move out into the garden. This schedule works nicely with one small, heated, germination cabinet.

The germination cabinet

This priceless tool is merely a container roughly the size of a cardboard apple box from the supermarket, whose temperature can be held warm and stable. Unless there is a place in your house where the temperature in springtime is always over 75°F (24°C), all day, and all night, then for predictable results make a germ box. Don't be daunted. A germ box is the easiest thing imaginable to make or improvise. And once you've got one you'll use it for a few other things that'll make your gardening a lot more reliable. (It can also be a good place to raise a loaf of bread.)

Few gardeners will want to start more than a dozen plants at one time. A batch of seedling pots won't be in the germ box for more than one week, so during the spring it can be used to start several batches in sequence.

For many years the oven in our kitchen stove was my germ cabinet because it is a well-insulated box with a 25-watt bulb installed. With the light on, the oven temperature stays about 75°F. After my wife, Muriel, repeatedly objected to this use of *her* stove in *her* kitchen, I began to use a germ box made of scrap wood with a small lightbulb on the bottom. It has a sliding glass top (cut from a scrap of double-strength window glass, it sits in grooves or tracks) so that by a combination of changing lightbulbs from 25 to 40 watts and/or opening the top a few inches, I can control the temperature inside to suit. A cardboard apple case would also do well.

You will need to calibrate your box — in other words, figure out how to get it to hold at 75°F to 80°F (24°C to 27°C) by either changing bulb size or adjusting the air flow. Beware: Temperatures exceeding 80°F may lessen germination. Also beware of the fire risk in a cardboard box when you use bulbs larger than 25 watts.

Procedure

The sealed pots can rest on an old cookie sheet or on a rack. For more uniform heat distribution, they are best supported above the lightbulb rather than below it. Three days after sowing, begin checking each pot twice a day. As soon as some of the seedlings have emerged in any pot, remove that one from the germ cabinet, remove the nearly airtight plastic bag, put the pot on a growing tray or shelf in your brightest, most sunny window, and begin watering it as needed. Strong vigorous tomato, pepper, eggplant, and cucurbit seeds should appear within four days; any seeds that take an entire week to emerge will be spindly and lacking vigor and will have difficulty getting going.

When the seedlings have fully opened their first pair of leaves (cotlydons), use a small pair of scissors or your fingernails to cut off all but the best three seedlings per pot. When the seedlings have developed two true leaves, thin the pot down to two seedlings. When the seedlings have three true leaves, cut off one, leaving one to grow.

When you have thinned down to one seedling per pot, put that seedling outside in full sun during the day when the weather is pleasant. If possible, for the last week before transplanting, keep it outside overnight except when it'll be shockingly cold or when frost threatens.

Fertilizer

Raising transplants is an occasion to use liquid fertilizer. If you make the potting soil rich enough to push the seedling as fast as possible, the high level of soil nutrients will also encourage soil diseases that attack emerging seedlings. From the time you sow until about the time the first true leaf forms, it is best if the seedling mix provides only minimal NPK (remember: nitrogen-phosphorus-potassium) but supplies plenty of calcium because this essential nutrient can't be added conveniently in a liquid fertilizer. When the seedlings have developed enough strength that they can resist fungal diseases (for most vegetables, this is when they have developed one true leaf) it is time to start pushing them. For this purpose, add fertilizer to their water.

Organic liquid fertilizers. Organic gardeners, I apologize; the most effective liquid fertilizers are not organic. That is because it is nearly impossible to get organic sources of phosphorus to go into solution. Liquid fish emulsion and liquid seaweed preparations are deficient in phosphorus. But seedlings need a lot of P if they are to become stocky and strong. If you find a liquid concentrate that says it is organic and also provides a high level of phosphorus, it is almost certain the maker has fortified some fish emulsion with phosphoric acid. Now, phosphoric acid is a most effective fertilizer; it just doesn't meet the definition of "organic" as prescribed by the certification authorities.

The best purchasable genuinely organic compromise is an equal mixture of fish emulsion and liquid seaweed. The fish is high in nitrogen; the seaweed has potassium, trace minerals, and assorted hormones that act like vitamins for a seedling. In combination they grow a reasonably healthy, reasonably stocky plant. Another thing that acts as pretty good fertilizer is strong cold brewed black coffee. Even better than the brew is the grounds. Used coffee grounds are a seedmeal that hot water has been passed through. Judging by

Small batches of seedlings can easily be raised in a sunny window, sitting on a table or shelf. If the light is not bright enough, if the seedlings incline strongly toward the light, or if their stalks seem too thin and spindly, there is a simple and almost costless method to significantly increase the light level. Make a reflector of a large piece of cardboard perhaps 14 inches (35 centimeters) tall and somewhat longer than the line of pots. Cover one side of it with aluminum foil (glue it on or wrap it around to the back and tape it down) or paint it bright white and then prop up that reflector immediately behind the seedlings.

If the seedlings still seem a bit leggy and weak-stemmed, they need exercise. Get a small window fan and make the airflow shake them gently, or open the window a bit. A few hours of gentle breezes every day will cause seedlings to increase the size and mechanical rigidity of their stems. ■

how coffee makes plants leap forward, I would reckon the grounds to be about half as strong as chicken manure. At times I've arranged to take away all the coffee grounds from a nearby restaurant that has a busy espresso machine. I spread them and dig them into any area of my garden that is being prepared for planting. They could also be put into a compost heap in place of animal manure.

Chemical liquid fertilizers. Any brands sold commonly in garden centers and supermarkets that are about 20-20-20 and also list trace minerals on the label work quite well. The best and most costly chemical fertilizers are hydroponic nutrient solution concentrates. Hydroponic fertilizers formulated to produce vegetative growth should also supply reasonable amounts of calcium, a difficult chemical feat.

Amount of fertilizer. Potting soil made with COF, as I suggested, grows healthy plants, but grows them slowly. When you try to speed results, there is a far greater risk from overfertilizing, which poisons the seedling, than from using too little. Even slight fertilizer toxicity will prevent seedlings from growing as well. This is particularly dangerous because the typical response to poor growth is to think the plant needs even more fertilizer. Take no chances! Mix liquid fertilizer, be it organic or chemical, at one third the recommended strength and then use it at that dilution about every other time you water. Seedlings mainly need fertilizer when the sun shines on them and makes them grow, so during cloudy spells, when plants don't grow much, you naturally don't water them much. When the sun is shining and you're watering more frequently, the plants will automatically get the increased fertility they need to grow.

Transplanting tips

By paying attention to the following tips, you can reduce transplanting shock and see your seedlings get off and growing quickly. After all, isn't that why you went to all the trouble of raising seedlings ... in order to harvest them as soon as possible?

Set each seedling into a super-fertile hill. After preparing the bed or row, remove one big shovelful of soil from each spot a transplant will go. Set the soil beside the small hole you've made. Into the bottom of that hole toss half a cup (125 milliliters) of COF or about a pint (500 milliliters) of strong compost.

By digging, blend the COF or compost into about a gallon (four liters) of soil so that most of this amendment will be located below the seedling.

Now slide the loose earth you removed at first back into the hole. Smooth out the spot and scoop a small hole in the center with your hand. The hole should be about twice the volume of your seedling's rootball and deep enough that when the seedling is put in and the hole refilled, the soil line will fall just below the seedling's first two leaves.

Unpot the seedling and gently place the rootball into the hole, minimizing damage to the roots and avoiding breaking off any soil from the rootball.

Mix a bucketful of full-strength liquid fertilizer or use compost/manure tea. Take a quart jar or tin can, scoop out a quart (liter) of fertigation (see Chapter 6), and pour it into the hole, gently enough to not wash soil away from the rootball but rapidly enough to fill the hole before it soaks in. Then quickly push loose soil back in around the seedling's rootball, creating a muddy slurry that will settle into all the nooks and crannies of the rootball, causing a tight connection to form between the roots and the surrounding soil. If you do this step right, the seedling will not need watering any more often than anything else in the garden from then on. It'll grow really fast, too.

Reasons not to use transplants

For gardeners operating in a short-season area, setting out transplants is the only way to get more than a few ripe tomatoes before the summer is over. Even if you enjoy a long frost-free season, getting your first ripe tomato six to eight weeks early is a worthy ambition. The same is true for peppers and eggplants. So raising seedlings for these species and transplanting them out as early as possible is not a bad idea.

But when it comes to other vegetables, using seedlings doesn't make as much sense.

You don't save all that much growing time. No matter how skillfully you set out the transplant, no matter how well hardened-off the seedling is, transplanting almost always sets back growth. A five-week-old seedling set out at the same time that seeds are sown will usually come to harvest only about three weeks ahead. And you can often get spring crops to germinate earlier, in colder and less-hospitable soil than you might think they could, simply by "chitting" them before sowing. (Chitting will be fully explained in Chapter 5.)

So instead of feeling urgency in spring, relax. Accept that directly seeded crops will start yielding a few weeks later than they might had you used seedlings.

Suppose you want a continuous supply of small cabbages, two each week. Suppose, where you garden, the earliest that cabbage seeds can sprout outside is three weeks after the spring equinox. On that date you sow four clumps of cabbage seeds directly in their growing positions, enough to supply the kitchen for two weeks. At the same time you transplant six tough, well-hardened-off seedlings that won't go into shock when hit with a light frost. Even though these will be set back a bit when you put them out, they'll still mature a month ahead of the direct-seeded cabbages. To maintain a steady, continuous supply, directly seed four more plants every two weeks from that time onward.

Consider a slow-growing crop like celery. Raising seedlings to transplanting size takes not five weeks, but ten. Two and a half months of tending little plants in trays seems a bit much to me. Instead, I sow celery (and celeriac) directly in place, but in my mild winter maritime climate I do it rather late in spring so these tiny seeds have time to germinate before hot weather comes. Where winters mean freezing soil, directly seed celery in mid-spring for late-summer and autumn harvest. What's wrong with that? There'll be plenty of other things to eat during high summer. Where winter is entirely mild, you can direct-seed celery after the heat of summer has passed and it becomes a cool-season or winter crop. Whatever you have to do to insure that celery seed germinates, it will end up less trouble than tending seedlings for ten weeks and then transplanting them.

Direct seeding produces a stronger root system. Many vegetable species make a taproot. But when the young plant is confined to a small pot, the taproot disappears. The transplanted seedling then forms a shallow root system lacking subsoil penetration. This makes little difference as long as you have plenty of irrigation. But should you try to grow a garden on rainfall or in less than ideally fertile soil, the vegetables' ability to forage in the subsoil is critical.

Don't be a slave to your garden. Growing transplants takes a lot of close attention. They have to be watered every day that it isn't raining, and sometimes, if you use small containers so as to get a great many of them going in a small space, you have to water twice a day. Neglect them just once, go off on a visit and forget to come home in time one sunny afternoon, and the whole effort of weeks might be lost in one mass wilting. Seeds sown in the same place they'll grow to maturity are far more capable of taking care of themselves.

The garden center seedrack

If you're taking my advice and avoiding garden center transplants, you may turn instead to the rack of seed packets on the other side of the store. But now I'll tell you why you shouldn't buy most of those either.

When I began gardening in the 1970s, seedrack picture packets were low-quality stuff. I found a broader assortment of varieties by mail order, but most of it proved to be the same low quality. There were also a few ethical mail-order seed companies. There still are, but 30 years later the overall situation is even worse.

WalMartization has further degraded the picture-packet rack. Independent garden centers, unable to compete, are vanishing. Transnational retail chains use enormous buying clout to squeeze seedrack jobbers' profits. In consequence, in the same way there have come to be fewer retailers, there now are fewer large seedrack picture-packet companies, and the survivors are struggling ever harder to turn an ever-dwindling profit. So, in turn, the desperate seedrack companies have no choice but to demand even lower prices from *their* already low-priced suppliers, companies specializing in growing cheap garden seed for seedrack and mail-order retailers. These suppliers, the actual producers of seed, are termed "primary growers."

In my first business I learned that every product or service could be compared to a three-legged stool, with the legs being price, quality, and service. If you lowered one leg, you had to lower the others accordingly or the stool tilted. Cut price, and the quality and service have to go down similarly. Ask for an increase in quality or service and you have to be prepared to pay more. So the primary growers of cheap garden seed, faced with an implacable demand for even lower prices, cut quality. They had no choice.

How do you cut quality on something that is already low quality? You're about to learn a few things about how the garden-seed business really works.

Sweepings off the seedroom floor

The year I put out my first mail-order catalog, I visited the nearby regional facilities of a primary seed grower. This company produced most of the traditional well-known open-pollinated varieties used in the garden-seed trade. Its garden-seed prices were extraordinarily low. If I wanted to offer the traditional varieties familiar to most gardeners, I could find no other source to buy them from.

The district manager was pleased to meet this newbie to the trade, share a cuppa, and see what he could do to more firmly cement a relationship. In a fatherly way he set out to educate me. He explained that I should continue to buy most of my seed from his company because it supplied most of the seed used by most of the mail-order catalog companies and picture-packet seedrack jobbers in the United States and Canada. Because it did such high volume, it grew new seeds for most of the varieties every year. Thus I would not be getting old seed from him, no weak stuff that had been sitting around the warehouse for years before it became "new to me."

Then he informed me that to please the home gardener, the most important thing was that the seeds I sold came up. If after that the seeds did not grow too uniformly and did not produce the highest-quality vegetables, it did not matter. Of course, he said, at its low, low prices, his company's garden seed could not be of commercial quality. "But," he said, "the gardener is not a critical trade. You can sell the gardener the sweepings off the seedroom floor."

Sweepings? When seed is harvested it is mixed with chaff, soil, dead insects, weed seeds, and other bits of plant material, so it passes through machinery that shakes and blows and sifts and separates in all sorts of clever ways. The good fat dense seed goes into the bag; the chaff, soil, small seed, weak seed, light seed, and immature seed tends to fall to the floor. This is the "sweepings." In this case the term was a metaphor; he was referring to other aspects of quality.

Critical trade? The farmer or market gardener is a critical trade. Suppose the crop is cabbage. If it seems to grow okay but a large percentage of the plants fail to head properly, farmers know they've been defrauded. There may be a lawsuit because they will have lost a huge sum. Farmers know how many days it takes for a familiar type of vegetable seed to come up. If broccoli seed usually emerges in four to six days, but the lot just planted took eight days to show and the weather wasn't cold during those days and the emerging rows are spotty, the seedlings look spindly, and many of them proceed to die before they get established, farmers have no doubt that the seed was weak. At very least they are going to be looking for another seed supplier and will grizzle about Company X at the coffee shop. But let that same thing happen to gardeners, let seed germinate badly or fail to yield uniformly and productively, and inexperienced gardeners, not being a critical trade, wonder if it was their watering, their soil preparation, the depth they sowed at, or any of a handful of factors they are uncertain about. Almost never does the home gardener blame the seed. Besides, the gardeners are out only a few dollars for a small packet, where farmers might have invested a few hundred dollars per acre for a field of 80 acres. And probably each of those 80 acres consumed a few hundred dollars' worth of tractor work. When farmers take a $16,000 loss on seed, they don't just shrug.

What could make a worse loss than a poor germination? Suppose the seed came up vigorously and grew rapidly, but at maturity yielded something surprising, something unmarketable. In that case the unfortunate farmers would have spent additional hundreds of dollars per acre on pesticides, cultivation, etc. Not to mention what lawyers call "lost opportunity." Yes, those farmers wouldn't just shrug.

Ah, ha! said I, the novice seedsman, to myself. Too bad I'm committed to buy from this guy for this season. Before I order any more seed for next year, I am going to find high-quality suppliers! And it also became clear to me at that instant that the only way to be sure of what I'd be selling would be to put a lot of effort into my variety trials.

Commercial quality seed

Two factors make seed suitable for market gardening and farming — and highly desirable for home gardening, too: genetic uniformity and vigor.

Genetic uniformity. When you sow seeds for iceberg lettuce, you want a row that makes heads that will hold for a week or more without becoming bitter. But suppose a large percentage of those plants emit bitter seedstalks *before* the head is ready to be cut. If you're growing broccoli you want to cut flavorsome, small-beaded, central flowers without a lot of small leaves coming out of them, not a loose flower with enormous (coarse), harsh-flavored beads that are already turning yellow and preparing to open before the flower has reached half its final size.

Most varieties are prone to deviation, especially so in species whose pollination is done by insects or wind. What stands between the variety's ongoing usefulness and its deterioration is the plant breeder, who spots chance mutations and unintended crosses with other varieties in the seed production field and gets rid of them. Every seed production field must be patrolled regularly, with each and every seedmaking plant studied carefully to make sure that any off-types are removed. If the species is insect- or wind-pollinated, someone must also patrol the surrounding area to make sure that no flowering member of the same species is growing in some neglected field or someone's home garden. To maintain a quality variety takes a lot of skilled work. That seed has to be sold at a rather high price.

It is easy to produce cheap seed. You simply do not bother eradicating off-types, do not have a highly paid plant breeder doing magical tricks to purify lines, do not rigorously patrol the fields and gardens surrounding production areas, do not carefully hand-select the plants that are allowed to make seed. You plant the seed production field, let it grow unsupervised, harvest the seed, and the next time you are about to grow that variety, you dip into the bag you grew the last time and use it to plant the next year's production field. When done this easy way, each successive generation becomes ever more variable and ever less productive.

If the variety gets too degraded, the bargain-price primary grower may buy a few pounds of expensive commercial-quality seed of a different but similar-looking variety from a quality seed company and use that to start the next seed production field, whose harvest will be (mis)labeled with a well-known heirloom name. This is why a lot of the "heirlooms" produced by low-price primary growers have nothing in common with the original variety for which they were named.

What I have been describing was standard practice in the cheap end of the home-garden seed trade long before I went into it. But when seedrack jobbers demand even lower prices, the primary garden-seed grower can no longer afford to start out with decent seed.

Vigor. Newly harvested seed usually starts out germinating strongly. Over time, every lot will eventually become so weak that the few sprouts still able to emerge from their seed coats in a sterile germ lab could not possibly establish themselves under real conditions — in the field. Some species have a short storage life. Parsnip seed, for example, rarely sprouts effectively in the third year from date of harvest. The cabbage and beet families routinely sprout acceptably for seven years.

Savvy market gardeners and farmers require a minimum level of laboratory germination for each vegetable species. Should germination fall below that figure, it means the seed is almost certainly too weak to depend on. Farmers and market gardeners find out what the germination of a seed seller's lot is before they agree to accept it. Territorial Seed Company wouldn't buy cabbage seed unless it was 85 percent germ or better. When I was offered a lot exceeding 90 percent I would smile because it was a virtual certainty that it would sprout strongly. If I had to carry over unsold seed from that lot, it would certainly sprout well the next year, too, and would likely sprout well enough to sell proudly two years later.

Because germination ability is a major question when buying seeds, and can become a thorny legal issue between buyer and seller, there is a body of seed law. The rules of Canada, the United States, and the European Union are quite similar. In the US, for example, any time a package of seed weighing more than one pound (454 grams) is sold, it must show the results of a germination test done at a certified laboratory within the previous six months.

If home-garden seed packets had to be labeled this way, it would make them cost a few cents more per packet. Mainly to create an illusion that the small user of seeds is protected without having the germination stamped on every packet, a regulation termed "USDA Minimum Standard Germination" (or "Canada Number 1 rules" or "EU Minimums") was established. This regulation includes a long list of minimum germination levels for each vegetable species (see Figure 4.2). If the seed sprouts below that figure, it may not be sold unless the packet is plainly marked "BELOW STANDARD GERMINATION." But if the seed is at or above minimum germ, nothing need be said; it is assumed the seed is okay.

I said "illusion" in the previous paragraph because the minimums were created by the seed industry for its own benefit. They are so low that no critical buyer would ever knowingly purchase seed that was anywhere near the minimum germination levels.

During the 1980s, Johnny's Selected Seeds (an admirable American mail-order seed company) included in its catalog a chart showing the first two columns in Figure 4.2. Rob Johnston, the company's owner, would not knowingly sell seeds that were unlikely to come up and grow in the field, even though it was legal to do so. Please think about what Rob had to go through when buying seeds himself. He would have to bring seeds into his warehouse at germination levels significantly higher than his sell-at minimum. That was the only way to insure they would still be above his minimums six to nine months later, when the user put them into the earth. When I ran Territorial Seeds I tried hard to buy seeds at levels at least 5 percent above Johnny's minimum levels.

Germination standards

	USDA minimum standard germination %	Johnny's Selected Seeds minimum germ at time of sale %	Commercial quality minimum germ % *	Storage potential in years % **
Beans, snap and dry	75	80	85	3
Beans, lima	70	75	80	3
Beet, Swiss chard (beetroot, silverbeet)	65	75	80	6
Broccoli	75	80	85	6
Brussels sprouts	70	75	85	6
Cabbage	75	75	85	6
Carrot	55	70	75	6
Cauliflower	75	75	85	6
Celery	55	60	65	5
Chinese cabbage	75	85	85	5
Collards	80	80	85	6
Corn	75	80	85	3
Cucumber	80	80	85	6
Eggplant (aubergine)	60	70	80	5
Endive	70	70	80	5
Kale	75	75	85	6
Kohlrabi	75	75	85	6 ☞

Figure 4.2

When a seed seller wishes to cut costs, one way is to make no effort to have its packets exceed minimum standard germination. That policy will not mean that all the packets on the rack will germinate poorly. It does, however, mean that not all of the packets will germinate well. And if the company wishes to be dishonest as well as unethical, then it may make no effort to insure its seeds are above minimum standard germ, and hope their state Department of

	USDA minimum standard germination %	Johnny's Selected Seeds minimum germ at time of sale %	Commercial quality minimum germ % *	Storage potential in years % **
Leek	60	70	75	3
Lettuce	80	80	85	4
Muskmelon	75	80	85	6
Onion	70	75	75	3
Pak Choi	75	80	85	6
Parsley	60	60	70	6
Parsnip	60	60	70	2
Pea	80	80	85	3
Pepper	55	70	80	4
Pumpkin	75	80	85	6
Radish	75	80	85	6
Rutabaga (swede)	75	80	85	6
Squash	75	75	85	6
Tomato	75	80	85	4
Turnip	80	85	85	6
Watermelon	70	80	85	6

* These figures are my personal best guess of what a knowledgeable and skillful market gardener or farmer would wish for. ** This data pertains to Chapter 5. It is included here to save space. ■

Agriculture seed cop doesn't spot this and slap its wrist with a minuscule fine — which is all that happens.

Is it any wonder that people who buy picture-packet seeds come to believe that they can't reliably make seeds come up and choose to use transplants?

Regionality

When a seedrack jobber operates across the continent and seeks to cut costs in every possible way, one of the things that tends to go by the boards is any attempt to offer regionally appropriate varieties. The compromise is to try to come up with an assortment that appears to work everywhere. This cannot be done, as I will explain in the next chapter.

CHAPTER 5

Seeds

I have just trashed a large percentage of garden centers. Some sell transplants whose use frequently leads to less-than-ideal results. Most sell seedrack picture packets that offer less than ideal germination, and more than a few of these packets contain poorly selected, nonproductive varieties or varieties not at all adapted to the climate. However, there still remain some independent garden centers run by people who know what transplant quality is, who sell seedlings meeting a high standard, and who offer better-quality seed assortments from mail-order companies that also produce a few seedracks.

If your food gardening is little more than a backyard hobby, an amusement, an entertainment that leads to a random mix of positive outcomes and disappointments, then getting great seeds and seedlings is of little consequence. But for me, gardening has never been a minor affair. It is life itself. It is independence. It is health for my family. And for people going through hard times, a thriving veggie garden can be the difference between painful poverty and a much more pleasant existence.

Starting the garden with quality seeds is essential because of what I call "windows of opportunity," those brief periods when each crop may be started. The gardener tosses some seeds or seedlings through that window of opportunity and hopes they take root and grow. Then the window closes. For some vegetables there are repeated chances to start a crop. For others the window may only be open for a few weeks. Miss it and there will be no crop that year.

The seeds tossed through that window can take a week to ten days to emerge. There may be but one chance to resow if you immediately realize that the first sowing has failed. The worst disappointment occurs when poor seeds do germinate and seem to grow adequately, but yield little or nothing. The next worst happens when the seeds germinate a bit slowly and a bit thinly and then proceed to grow poorly due to lack of vigor. In this case, gardeners rarely realize what is happening in time to resow. Gardeners who depend on their gardens know they can't take chances; they grow their own seedlings and source quality seeds by mail.

The mail-order seed business

The word "ethics" means doing what would probably result in the greatest good for the greatest number of people. The opposite of ethics is criminality, which is about getting something for yourself without giving back anything in exchange. Many individuals are in business primarily for profit, for self. This low level of ethics produces little concern for other people, who are supposed to look after themselves in every transaction. This is the source of the old expression "Let the buyer beware." Other individuals take great joy in doing business primarily to provide a service to others, and secondarily to make a living.

Being ethical does not mean your responsibility is discharged merely by following the letter of the law. It means that your business does what is necessary to provide the service it was set up to accomplish.

If you were to ask the owners of an ethical vegetable-garden seed business what the goal and purpose of their business was, the response would be something like: "Our company exists to help independent people who grow their own food become more self-sufficient, healthier, and more economically secure. To do that, our business —

- provides seeds that have a high likelihood of germinating,
- provides varieties that are adapted to thrive in the buyer's climate,
- provides varieties that are suitable to a self-sufficient homestead lifestyle (instead of selling varieties that suit the needs of industrial agriculture), and
- honestly and accurately describes the performance and qualities of the varieties sold."

Of course there are additional (and entirely ethical) reasons to sell vegetable seeds. Some companies specialize in gourmet or unusual vegetables. This is a backyard hobby market in which the customer's main concern is not reliability or production. It could be perfectly okay to sell this sort of customer rare seeds that don't germinate terrifically well. It could also be okay to sell a gourmet variety that is irregular, with many individual plants proving to be nonproductive. Another sort of seed company specializes in heirlooms. Often its seeds are organically grown and produced by a network of collaborating amateurs. Because it is amateurs doing the production, their varieties often become irregular, inbred, weak. Because the enthusiast often has little idea of the correct way to handle and store seed, the germination levels tend to be sub-par. Not always, but frequently. Coping with these uncertainties doesn't matter to someone with a passion for knowing and growing antiques, but degraded heirloom varieties might not suit someone who wants to efficiently grow nutritious, chemical-free food.

To insure it is selling the types of seeds and varieties it wants, the owners of an ethical seed company will take the following steps to fulfill the four points listed above.

Germination

In the discussion of Figure 4.2 in Chapter 4, I mentioned that Johnny's Selected Seeds included information on the minimum germination level of the seeds it sold. Other mail-order seed companies also have in-house minimum germination levels significantly higher than the standard set out in the seed law and, as commercial suppliers do, perform germination tests twice a year on all lots of seed in their warehouses. This sort of company either has a large market gardener and farmer trade as well as home-garden customers and sells the same seedlots to both, and/or the company wishes to act with the highest possible level of responsibility.

It must seem easy to be slack, unethical; so many are. But it is not easy to send thousands of dollars of weak seed to the garbage every year. If you're only supplying the home gardener it can be tempting to sell weak stuff anyway. After all, the gardener is not a critical trade ...

Variety trials

To honestly describe its varieties, the seed company must actually grow trials. This is also the only responsible way to choose which varieties to sell. There is nothing particularly difficult about conducting trials; it just takes a bit of land, focused effort, and a fair bit of money. I reckon the minimum size for a meaningful trials ground would be about half an acre (2,000 square meters) for a small homestead seed business. A medium-sized mail-order company might use a few acres. I know of three large mail-order businesses using about ten acres (four hectares).

Trials are usually laid out in widely spaced rows so that each plant can be evaluated. Enough plants of each variety are grown to determine if the variety is uniform — to see if all the plants are equally productive. If it is a cut-once or yank-out vegetable like cabbage or heading lettuce or carrots, this question needs to be answered: Will the plants all mature at once or is the harvest period spread out? The commercial grower wants uniform maturity. The family kitchen usually prefers the harvest to be spread over a long period unless it intends to can or freeze the harvest. A worthwhile trial needs at least five or six plants for something like a cabbage or cauliflower variety, and a minimum of a ten-foot (three-meter) row for something like a carrot or beet variety. Thus, to do a cabbage trial, one might grow five plants each of 20 varieties being inspected, or a total of 100 cabbage plants. A trial like this can show which varieties are well suited to the home garden, how long they take to head up, how big they are, how long they hold before bursting, how much frost they'll tolerate, if they're tender enough for good slaw or tough enough for good kraut, etc.

Is there any other way to determine a catalog offering? Sure, but it would not be as ethical. A company can assume that commonly known varieties must grow well, or it could ask the local agricultural extension office what suits the area and sell that. A seedsperson could ask suppliers which varieties are good for the home gardener and then accept what is offered. But for real, believable, and accurate information, you have to do trials.

Only your own variety trial can reveal what tastes good. Reading between the lines of research-station trial reports won't help you find that. The main test I use in my own trials is to taste everything raw in the field. If a variety doesn't pass that test, it rarely progresses to an evaluation of its taste when cooked. If the trials-ground master doesn't like raw veggies, he or she will probably rate things differently. And that's what makes a horse race. But whether raw fooder or cooked fooder, I don't believe it's possible to have effective home-garden trials conducted by someone who is not a serious vegetableatarian. Otherwise, the varieties are being evaluated in much the same way that commercial varieties are ranked — by appearance, storage potential, ease of harvesting, etc. Not that these factors are unimportant. But for the home gardener they are secondary to flavor, culinary qualities, storage potential in a root cellar, etc.

Organically grown trials can show which varieties resist diseases or insects. I repeat, *if they are done organically.*

"Grow outs," taking a look at how the seeds purchased for resale actually grow, also happen on a trials ground. Except for a few mail-order businesses dealing in heirlooms, retail garden-seed suppliers are mainly distributors. They buy bulk seed and repackage it. There is no commodity more open to misrepresentation than vegetable seeds. A bag of cabbage seed can correctly cost $10 a pound wholesale or $250 a pound wholesale. The contents of both bags look identical — small round black seeds. The bag is labeled with a germination percentage determined within the previous six months. This can be easily and inexpensively checked by the buyer within one week of receiving the seeds, simply by counting out 100 seeds and sprouting them. But what those seeds will actually produce is another matter. Cabbage seed that costs $10/pound does not produce a row of identical heads. Many of the plants may not form a proper head at all. Cabbage seed that costs $250 per pound should yield a row in which every plant heads up perfectly and identically.

From time to time a primary grower has been known to "accidentally" mislabel a bag intended for the home-garden trade, sometimes enclosing suspect seed from a discontinued variety worth next to nothing in place of its finest item. (This never seems to happen to the grower's commercial customers.) On one occasion the bag I bought labeled "kale seed" turned out to contain some sort of fodder plant that was virtually inedible by humans. The only way to purge these items from one's inventory is to grow a small sample of all the seed lots purchased. At least this way the incorrect seed will not be sold for more than one season. The wronged seed merchants can then complain and demand a refund, keep their suppliers on their toes, and reduce the likelihood of being the recipient of such "mistakes" in the future.

If the retailer grows out everything in its catalog for which a new lot of seed was purchased that year, that would make a pretty large garden in itself. And doing it costs a pretty penny.

Adapted to the region

It would be profitable if one catalog could sell fine vegetable varieties that would grow anywhere. But that is not possible. To ethically sell across the whole English-speaking world, a company would have to operate a different trials ground in each broad climatic zone being served. Instead, mail-order suppliers tend to be regional. Unless, of course, they don't bother growing trials.

When you purchase seeds, you have a far higher likelihood of a successful result if the supplier's trials grounds are located in roughly the same climatic zone as your garden. Here is a "climate map" drawn in the broadest of verbal brush strokes:

Short-season climates. This area comprises the northern tier of states in the United States and that part of southern Canada within a few hundred miles of the U.S. border (the area of Canada in which over 90 percent of its citizens live).

Moderate climates. The middle American states, the east coast of Australia roughly south of Sydney, and the North Island of New Zealand are moderate climates. In the United States, this is where summer gets hot and steamy (the frost-free growing season is more than 120 days), and the winter is severe enough to actually freeze the soil solid at least 12 inches (30 centimeters) deep. This level of winter cold isn't felt Down Under except at higher elevations. To roughly delineate this area in North America, draw east/west lines from about the northern border of Pennsylvania and the southern border of North Carolina extending to the Rockies. Lower New York state, the part south of a line from Albany to Buffalo, might also be included. Maybe Connecticut, too.

Warm climates. This includes the southern American states and coastal Australia from Sydney north up to, say, Bundaberg. The soil here never freezes solid; the summers are long and hot. The climate may be humid or arid. The comparatively brief winters can occasionally be frosty but are mild enough to allow for winter gardening without requiring protection under glass or plastic.

Maritime climates. In North America this bioregion is sometimes called Cascadia. It includes the redwoods of northern California, extends into Oregon, Washington, and the Lower Mainland and islands of British Columbia, always west of the Cascade Mountains. England, Ireland, Wales, southern

Scotland, southern coastal Victoria, Tasmania, and the South Island of New Zealand have about the same climate. These regions usually have relatively cool summers. Rarely does the soil freeze solid in winter except at higher elevations and where it is isolated from the ocean's moderating influence. When a period of sub-freezing weather does happen, the earth doesn't freeze deeply, nor does the freeze last long. Winter gardening ranges from difficult to easy and productive.

Who to buy from

In 1989 I wrote an article for *Harrowsmith*, then a brave country-lifestyle magazine. I explained the garden seed trade and evaluated and ranked mail-order companies. Why do I say "brave"? Because mail-order seed sellers made up a large portion of *Harrowsmith*'s advertisers, and my article offended more than a few of them. First I sent out 69 questionnaires on *Harrowsmith*'s stationery, stating that I was the ex-owner of Territorial Seed Company, that I was writing an article evaluating garden seed companies, and that I might telephone for further information after the questionnaire's answers had been received. Some of the questions were: Do you have a trials ground? If so, how large is it? Do you have your own in-house germination laboratory, even if uncertified? If so, how often do you test the seed lots on your shelves? What germination standards do you use to decide if a lot is fit to sell? What percentage (or how many) varieties in your catalog are actually grown by your company?

After eliminating those who elected not to respond (about half, which was not a surprise to me), I then removed from consideration those without trials grounds. Out of 69, only 20 were left. After a probing telephone chat with the management of these companies, I found 11 were worth recommending. For this book I have expanded my recommended list to include admirable suppliers in the U.K. and Australia.

The following are the businesses (organized by climatic zone) I'd advise serious food gardeners to use for the essential core of their gardens. They are the companies most likely to supply first-class seeds. When you garden to produce a significant amount of your family's food, you can't afford to use poor seed!

Your current seed supplier may not be included in my recommended list. Why am I so critical? Because when I grew proper variety trials for eight years,

I saw undeniable and large differences between varieties. I want you to have the best possible chance of realizing a productive and useful outcome. I reckon you can't afford to experience anything else.

Short-season climates

Stokes Seeds. Stokes, a Canadian company located near Niagara Falls, has a 10-acre (four-hectare) trials ground open to the public, as well as an additional 24 acres (ten hectares) that is not open and is used for research, plant breeding, and seed production. The company's main income is from farmers and market gardeners, but the home gardeners get the same quality of seed as the commercials. I have never purchased a Stokes packet that failed to come up acceptably. If Stokes' catalog has any weaknesses, they will be found in its offerings of home-gardener-only vegetable species — items that have no commercial interest such as kohlrabi, kale, winter radishes, celeriac, and other unusual veggies. For these I often find Johnny's a better source. Stokes makes two identical catalogs: one in US funds; the other in Canadian.

Stokes Seeds, PO Box 548, Buffalo, NY 14240 USA.

Stokes Seeds, PO Box 10, Thorold, ON L2V 5E9 Canada.

Both countries served by 1-800-396-9238 and www.stokeseeds.com

Johnny's Selected Seeds. In 1973, Rob Johnston, age 22, bootstrapped Johnny's with $500 in operating capital. The company is currently using about 40 organically certified acres (16 hectares) for trials, research/development, breeding, and seed production. Johnny's has bred and now grows seed for a significant number of its own varieties. Johnny's catalog offers many organically grown items and avoids selling fungicide-treated seed. Like Stokes, Johnny's offers small packets suitable for home gardeners and also sells the same varieties in larger amounts to market growers. Besides a broad line of vegetable seeds, green manures, and cereals, Johnny's sells a wide choice of organically grown, state-of-Maine-certified seed potatoes, a range of garlic varieties, etc. The company routinely ships to Canadians; non-commercial quantities do not require special certification nor do they incur import costs. But Johnny's may not send fava beans to Canada (nor corn to B.C.). There are also Canadian import restrictions on living plant materials like seed potatoes and garlic, and on some grains. Canadians venturing beyond vegetable seeds would be wise to telephone or check Johnny's website before ordering. Johnny's ships overseas (to me), too.

Johnny's Selected Seeds, 955 Benton Ave., Winslow, ME 04901 USA; (207) 861-3900; www.johnnyseeds.com

Veseys Seeds. Veseys was started nearly 70 years ago by a market gardener who began importing high-quality European seeds for his neighbors. The company's varieties all pass the test of reliable short-season maturity in trials on Prince Edward Island, and a rigorous screen it is, too. Seed-germination standards are constantly monitored in the company's own germination lab. Veseys will ship overseas.

Veseys Seeds, PO Box 9000, Charlottetown, PE C1A 8K6 Canada; 1-800-363-7333; www.veseys.com/

William Dam Seeds. William Dam has just published its 56th annual catalog. The company operates five acres (two hectares) of vegetable (and flower) trials and focuses on high-grade hybrid Dutch imports. I am pleased to see many of my favorite short-season varieties in Dam's catalog. The catalog also contains some old and highly degraded open-pollinated varieties (OPs) offered for the "organic" trade that I wouldn't touch: specifically DeCicco and Waltham 29 broccoli, as well as Snowball cauliflower. Dam will ship overseas.

William Dam Seeds, PO Box 8400, Dundas, ON L9H 6M1 Canada; (905) 628-6641; www.damseeds.com/

Moderate climates

Stokes Seeds. Stokes' location straddles the short-season/middle-states line. For that reason this company's offering is appropriate for middle-states gardeners. However, the catalog lacks some of those really heat-loving species and varieties considered "southern" vegetables. Contact details above.

Johnny's Selected Seeds. Generally, varieties that grow well where summer is cool will do even better where it is warmer. Almost everything in Johnny's catalog should suit until you go far enough south that the summers get really steamy. My doubts about using these seeds south of Pennsylvania/the Ohio River are unproven, but likely correct. Contact details above.

Harris Seeds. A decade ago, after 100 years of family management, the last of the Harrises sold the business to an international agribusiness company. Not too many years later, some of the original employees bought back the business and it is again in private hands. Harris's main business is with farmers and market gardeners. Unlike Stokes, which sells its entire line to all customers, Harris has two catalogs and offers a somewhat limited assortment of the same quality of seeds to home gardeners. Because its trials grounds and breeding focus are on a climate much like that in which Stokes operates, the main reason to deal with Harris is to access some of its historic varieties, such as incredibly delicious Sweet Meat squash and the Harris Model parsnip. The catalog is definitely worth a look.

Harris Seeds, PO Box 24966, Rochester, NY 14624 USA; 1-800-514-4441; www.gardeners.harrisseeds.com/

King Seeds. Down Under gardeners will appreciate this broad offering of generally high-quality vegetable seeds from major international seed growers, as well as herbs (and flowers). The catalog is especially valuable for its many lettuce varieties not otherwise found in Australia. The only obvious weakness in the catalog is the low quality offered in a few types of vegetables the company probably considers unimportant, such as kohlrabi and Brussels sprouts.

King Seeds, PO Box 283, Katikati 3063 New Zealand; (07) 549 3409.

King Seeds, PO Box 975, Pentrith, NSW 2751 Australia; (02) 4776 1493. www.kingsseeds.com.nz

Southern Exposure Seed Exchange. SESE is owned and operated by an "egalitarian income-sharing community," whose main concern is safeguarding and advancing seed production that is not in the hands of global megabusiness. Were I gardening in the middle states of the US. I would use its offerings for, as its name suggests, it provides access to heat-loving, traditional varieties. SESE is hoeing a difficult ideological row as it tries to maintain a quality offering while (1) avoiding seeds that aren't organically grown, and (2) having at least 40 percent of its varieties produced by a network of relatively inexperienced small growers. Most of these, but not all, are certified organic, but their results are not always reliable. Many SESE varieties are family-preserved heirlooms. In my own garden I am not willing to waste valuable space and effort growing a low- or non-producing variety in an attempt to be on the side of the angels. I would not purchase the following refined types from them because these need the best-possible breeding to be productive: Brussels sprouts, cabbage, cauliflower, OP broccoli, celeriac, kohlrabi, Florence fennel, bulbing onions (their scallions will be fine), salad radishes, turnips, rutabagas (swedes). They sell top-quality organically grown seed potatoes that are certified virus-free by the state of Maine.

Southern Exposure Seed Exchange, PO Box 460, Mineral, VA 23117 USA; (540) 894 9480; www.southernexposure.com/

Warm climates

Park Seed Company. I smile when I read Park's catalog. Virtually every variety the company offers represents the finest breeding attainable, entirely appropriate to its almost semi-tropical climate. This regionality is the result of many years of rigorous trials. Park bulbing onions are plainly labeled as medium- or short-day. Park sells doubly certified seed potatoes (Irish), both organically grown and produced from virus-free tissue-culture clones. That means they'll be maximally productive. There are also seven varieties of sweet-potato shoots. Park's rigorous germination standards exceed federal minimums. Because of the hot, humid weather in South Carolina, the company's seed is stored in climate-controlled conditions, and all its small seed is packaged in moisture-proof foils. And Park's prices are entirely reasonable. It's as good as it gets. Park will ship to overseas customers and send catalogs to foreigners who request one by post. Its website is not set up to accept orders from outside North America. Those living up the east coast of Australia north from Kangaroo Valley, or on New Zealand's North Island, should take a look.

Park Seed Company, 1 Parkton Ave., Greenwood, SC 29647 USA; 1-800-213-0076; www.parkseed.com/

Maritime climates

Territorial Seeds. I opened the doors of this business in 1980, lifted it by its bootstraps, and sold it to Tom and Julie Johns at the end of its 1985 season. I have no ownership or other financial interest in TSCo now, so when I give it good marks, there is no conflict of interest. However, I admire Tom and Julie for growing the small business I sold them into a company as big as any in this list except Stokes.

The certified-organic trial and production fields exceed 40 acres (16 hectares), and all descriptions and days to maturity listed in the catalog are as experienced on the trials grounds. The company has a complete germination laboratory and a resident seed analyst. Its in-house standards exceed federal minimums, and the unpacketed seeds are stored in a climate-controlled warehouse (low temperatures and low humidity). TSCo is producing increasing amounts of organically grown seed. Its broad line of seed potatoes is both organically grown and certified disease-free by the state of Maine. Originally the focus of the company was to serve only Cascadian gardeners, but Territorial now has as many or more customers east of the Cascades as it does in western Oregon and Washington states. Non-Cascadians should be aware that some varieties on its website, eminently suited to winter gardening west of the Cascades, are too slow to mature before winter freezes the garden solid east of the Cascades.

Territorial Seed Company, PO Box 127, Cottage Grove, OR 97424 USA; 1-800-626-0866; www.territorial-seed.com/

West Coast Seeds. West Coast was started about 1982 by a Vancouverite, Mary Ballon, as the Canadian branch of Territorial. Now it is independent. Mary runs a large certified-organic trials ground on alluvial soil near Delta, BC, and otherwise runs her business much as Territorial does in the United States, albeit on a smaller scale. Lately, shipping through the biosecurity barriers to American customers has proved too difficult, and West Coast is abandoning its clientele south of the border. It will ship overseas.

West Coast Seed Company, 3925 64th Street, RR1, Delta, BC V4K 3N2 Canada; (604) 952-8820; www.westcoastseeds.com/

Chase. I have no complaints about any of the three fine U.K. seed houses — Thomson & Morgan, Suttons, and Chase — but Chase is my preference. It offers many highly desirable certified-organic varieties grown by two European quality seed houses, Rijk Zwaan and Sainte Marthe. As an indicator of its attitude consider this: Chase is the only company I know of on this planet that still offers reasonably uniform and productive OP Brussels sprout varieties. It is entirely happy to serve overseas customers and sends catalogs anywhere without a quibble. Chase is currently one of my mainstays.

The Organic Gardening Catalogue, Riverdene Business Park, Molesy Road, Hersham, Surrey KT12 4RG England; 01932-253666; www.OrganicCatalog.com/

New Gippsland Seeds is located in southern Victoria, Australia. The location is a bit warm to supply someone on Tasmania, but many of the company's varieties work acceptably here. Mainlanders as far north as Kangaroo Valley in New South Wales should be most pleased with the entire line.

New Gippsland Seeds, PO Box 1, Silvan, VIC 3795 Australia; (03) 9737 9560; www.newgipps.com.au/

Miscellaneous suppliers and sources worthy of note

Dave's Garden is a website providing handy access to nearly every source of seed and materials a North American gardener would want. www.davesgarden.com/

Fedco Seeds only sells from its extensive catalog (downloadable from its website) and only during the spring. These folks run an honest business. PO Box 520, Waterville, ME 04903 USA; (207) 873-7333; www.fedcoseeds.com/index.htm/

Green Harvest Organic Garden Supplies provides Australians with a broad assortment of natural pest-management materials, as well as seeds, books, tools, etc. 52/65 Kilkoy Lane, via Manley, Qld 4552 Australia; (07) 5494 4676; www.greenharvest.com.au/

Landreth, the oldest continuously existing American seed business (since 1796), long in decline, was recently purchased by enthusiastic new owners who are making huge improvements. The new Landreth holds promise for gardeners in moderate climates. 650 North Point Road, Baltimore, MD 21237 USA; 1-800-654-2407; www.landrethseeds.com/

Lockhart Seeds, mainly a supplier of farmers in California's central valley, offers a line of seeds particularly suitable to the Californian homesteader. PO Box 1361, Stockton, CA 95205 USA; (209) 466-4401; no website.

Nichols Garden Nursery, specializing in the unusual and gourmet, has long been in business. 1190 Old Salem Rd. NE, Albany, OR 973211 USA; (541) 928-9280; www.nicholsgardennursery.com/

Peace Seeds, a breeding service and organically grown gene pool for the Cascadia bioregion, is run by my friend Dr. Alan Kapuler. Request his astonishing seed list by post. 2385 SE Thompson St., Corvallis, OR 97333 USA.

Peaceful Valley Farm Supply does sell seeds, but it is listed here primarily for the broad assortment of holistic gardening supplies, organic pesticides, etc., that it offers. PO Box 2209, Grass Valley, CA 95945 USA; 1-888-784-1722; www.groworganic.com/

Renee's Seeds offers a limited assortment of high-quality, rare, cottage garden seeds chosen by Renee Shepherd, author, expert gourmet cook and gardener. Sales only by internet or seedracks. 7369 Zayante Road, Felton, CA 95018 USA; 1-888-880-7228; www.reneesgarden.com/

Ronnigers Potato Farm has been doing business from a remote valley in the Idaho panhandle for 25 years. It offers about 100 organically grown seed potato varieties as well as garlic and related items. Star Route, Moyie Springs, ID 83845 USA; (208) 267-7938; www.ronnigers.com/

Select is a full-line seed company offering only the highest-quality varieties in substantial amounts in foil-sealed packets at surprisingly low prices (denominated in Swiss francs). Its colorful catalogs are in French or German only, but letters and orders in English are comprehended. Select cheerfully sends catalogs anywhere. Contact: Wyss Samen & Pflanzen AG, Schachenweig 14c, CH 4528 Zuchwil-Solothurn, Switzerland; www.samen.ch/

Importing vegetable-garden seeds

American home gardeners cannot at this time affordably import small quantities of vegetable seeds due to security concerns. Imports now require costly permits and inspections, hurdles to jump whose cost is far greater than any possible benefit.

The **United Kingdom** is blessed with many excellent suppliers. I find it hard to imagine someone living there having an interest in importing.

Canadians find importing much easier than Americans; however, their own companies do a fine job of serving their needs. Gardeners in the Lower Mainland of British Columbia might find the offerings from the U.K. to be of interest.

Australians can easily import veggie seed in quantities appropriate for the home garden without obtaining permits or following other expensive and awkward procedures — with the exception of four species: corn, beans, peas, and sunflowers. For all other veggies, all that is necessary is that the seed be in packets plainly marked with the correct Latin species name, a usual commercial procedure in any case. Importing is simple: order your seeds; pay by credit card; do not buy restricted species from overseas; await your seeds' safe arrival by post.

New Zealanders may import commercially sold vegetable seed, in packets appropriate to a home garden, for all species except beans, peas, corn, and beets. The seeds will be inspected by Ministry of Agriculture officials without any charge and released. Any questions should be referred to Ministry of Agriculture and Forestry Imports Management Office, PO Box 2526, Wellington 4-4989624 New Zealand. I have not tried to evaluate domestic New Zealand companies from Tasmania except to note that their prices are enormously higher than those found in North America, even figuring the exchange and postage.

Making seeds come up

A gardener who has difficulty getting seeds to come up is in a sad way, always needing expensive, unusual, and extraordinary solutions to a problem that shouldn't exist in the first place. The only conditions in which it should be difficult to sprout seeds are the still-chilled soils of early spring and the high heat of midsummer, when the earth can dry out rapidly. Achieving germination can also be tricky when you're growing veggies without any irrigation. However, even these stressful conditions can be surmounted with a bit of cleverness.

The best way I know to teach someone how to reliably sprout seeds is to tell them how a germ lab does it. Laboratory germination is accurate and duplicatable: two different germ labs testing samples from the same bag of seed should come up with the exact same result. To achieve uniformity of result, seed technicians have determined ideal sprouting conditions for each type of seed. Their procedure manual prescribes these ideal conditions, species by species.

Germ labs use sterile media (and the best kinds of sterile media), precisely controlled temperature, and precisely controlled moisture. Usually the test is done in a petri dish, a shallow, airtight, clear plastic container about four inches (ten centimeters) in diameter. For most kinds of seed, a technician will take a blotting paper disk slightly smaller than the inside of the petri dish, dip it into water, then squeeze all surplus water out of the blotting paper by balling it up in a fist and squeezing hard. The technician flattens the damp paper circle and places it in the bottom of the petri dish, counts out precisely 100 seeds, and places them atop the moist paper. He or she will then put the petri dish's cover on and place the dish in a heated box called a germination cabinet. Sometimes slightly better germination can be had by using a light, loose, finely textured, sterilized soil mix (usually consisting of sand, compost, and peat) instead of blotting paper. It is moistened to the same degree you would seek in the ready-to-till test described in Chapter 3.

The germ cabinet temperature is set to the optimum for the species under test. It could be as little as 60°F (15.5°C) for a plant like spinach to as much as 85°F (29.5°C) for okra or eggplant. Usually the temperature is set to be constant. Occasionally it changes regularly: 16 hours at a higher temperature, then 8 hours somewhat lower, much as soil would be when it is heated up with the sun shining on it and then cooled off at night.

The protocol for each species specifies how many days can be allowed for germination because any seeds that might emerge after a certain number of days would be too weak to survive real field conditions. The results of these tests almost always show that the batch that germinates the quickest also germinates at the highest percentage.

Field conditions are never as perfect as test conditions. Real soil is teeming with microorganisms, some of them hostile to seedlings. The temperature outdoors is never constant and often is colder than ideal, sometimes barely warm enough to let the seed get started. Sometimes the soil can get too hot. And soil moisture is never stable. So germination percentages in the field are never as high as the ones you get in the lab, but there is a relationship, and it

Germination in the field

Weather after sowing	Average soil temperature until emergence °F (°C)	95% lot percentage to sprout	Days to emergence	78% lot percentage to sprout	Days to emergence
Sunny, mild, and dry; watered alternate days	70° (21°)	60%	4½	25%	7
Sunny, mild, and dry; watered daily	67° (19.5°)	55%	5	20%	8
Overcast, mild, and dry; watered every three days	65° (18°)	58%	6	17%	9
Overcast, mild, and dry; watered every day	62° (16.5°)	53%	7	15%	9
Showery, cool	58° (14.5°)	50%	8	10%	10
Rainy, chilly	55° (12.5°)	40%	9	5%	12
Cold, showery	50° (10°)	20%	10	1%	14

Figure 5.1

is not what a mathematician would call "linear." What I mean is this: Suppose two lots of cabbage seed are sown at the same time in adjoining rows in the same garden at the same depth. Soil conditions are identical for both. One lot germinated at 95 percent in the laboratory, and all the seeds in that lot had fully developed (root and shoot and two small leaves) on the fourth day of the germ test. The second lot tested at 78 percent, and the last of these seeds weakly finished developing on the seventh day, the last day the protocol allowed. What typically happens in the field is shown in Figure 5.1.

Conclusion: To get the best outdoor germination, start with vigorous seeds and then assist them by creating soil conditions that match as closely as possible the conditions that have been determined to give the best results in a laboratory.

Watering less frequently

It's natural to fear that if seeds dry out while sprouting, they'll die. But every time we water, the soil temperature drops, slowing the seed's progress. Wet, cool soil enhances "damping-off," a fungus disease that invades and kills the seedling before and after emergence. When the stems of tiny seedlings pinch off at the soil line, you're seeing damping-off. This disease doesn't thrive in dryish soil.

Another kind of disease, powdery mildew, attacks members of the cucurbit family (squash, zucchini, melons, cucumber) in particular. Powdery mildew only thrives in damp, cool conditions. Most gardeners have seen powdery mildew cover the leaves of cucurbits at the end of summer when weather gets cool and humid. Powdery mildew also attacks cucurbit seedlings before emergence.

So it is wise to avoid watering sprouting seeds. But, you ask, don't seeds need watering?

Remember in Chapter 3 when I asked you to pay special attention to the moist footprints in a newly rototilled field? When soil is loosened up with shovel, tiller, or fork, its capillary channels are shattered. Until the soil resettles, moisture is not able to rise from the subsoil to the surface. Seeds dry out quickly in loose, recently worked soil. But what if, instead of cutting a furrow (sometimes called a "drill") with the corner of a hoe, we made the furrow by pressing the soil down, restoring capillarity immediately below the seeds. This

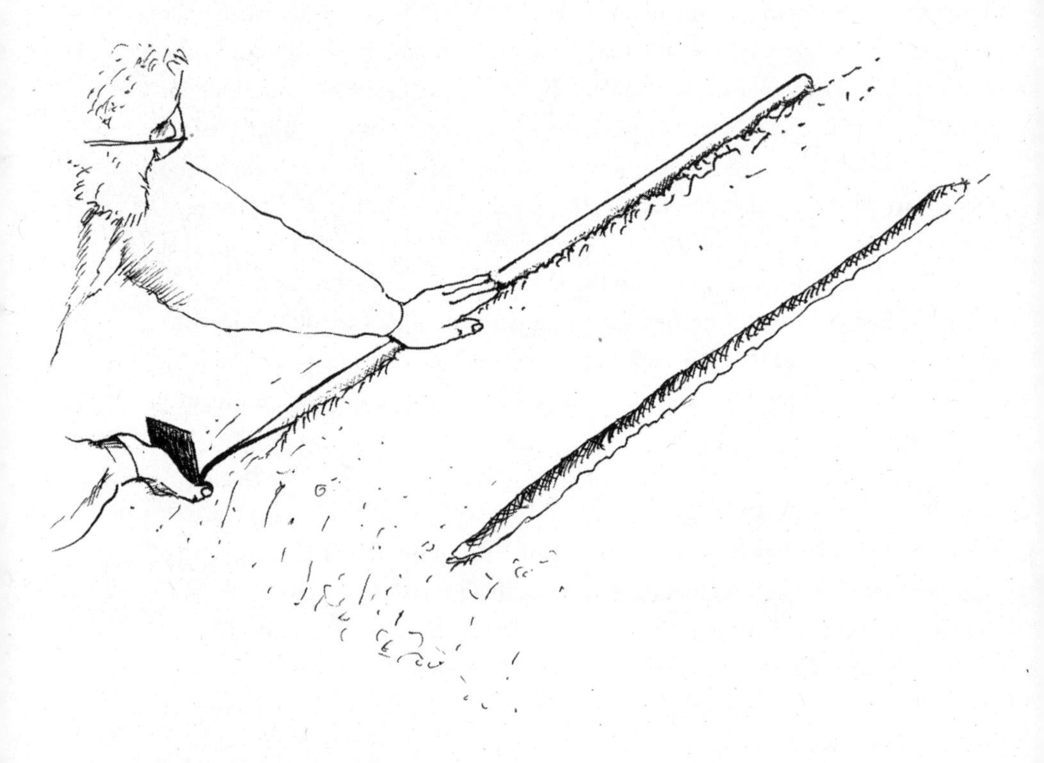

Figure 5.2: *Making a furrow by pressing down loose soil.*

is what I do whenever I sow small seeds close to the surface. I press the handle of my garden hoe down across the bed, making a U-shaped depression about half an inch to three quarters of an inch (1.25 to 2 centimeters) deep. I place the seeds at the bottom of this "U" and then cover them by pushing a bit of fine soil back into the depression. Now the seeds are sitting on top of their own version of a moist footprint, and the loose soil atop them acts to a degree as a mulch, reducing moisture loss. This method has another advantage: the entire furrow is the same depth, resulting in much higher and more uniform germination. If the bed was especially fluffed up, then before sowing I will give it a few days to resettle before pressing a furrow with my hoe's handle.

There is another step that'll also help the seeds. Recall my suggestion in Chapter 3 to rake in compost rather than digging it in, preventing crusting and puddling. This practice also makes the surface layer hold more moisture.

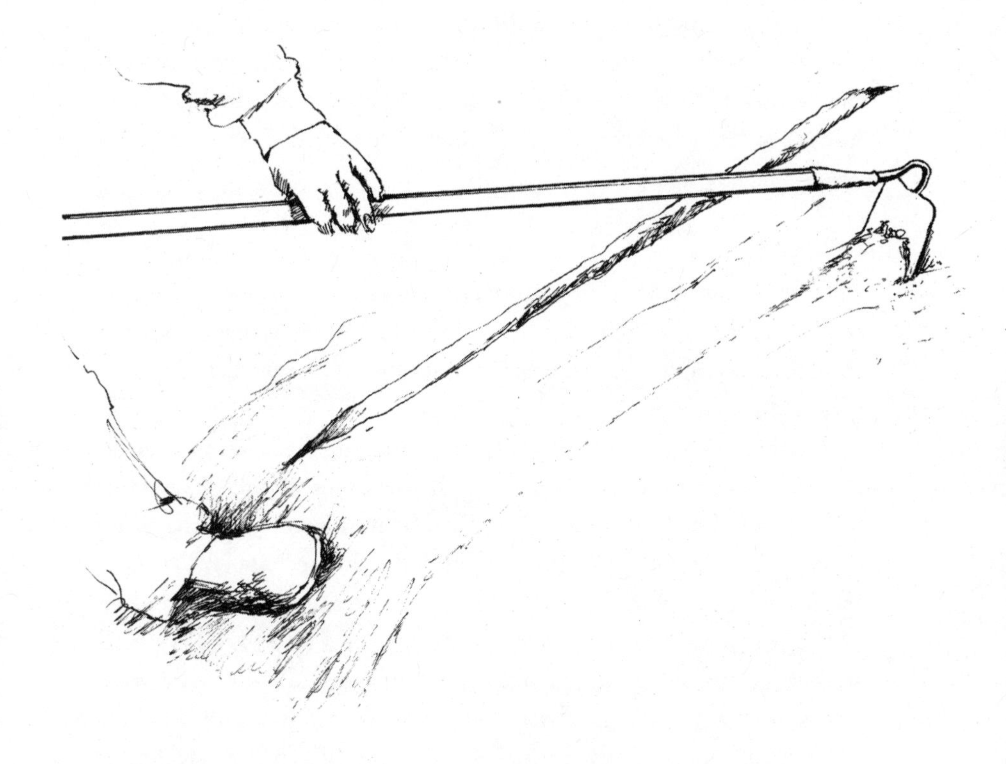

Figure 5.3: *Covering the seeds.*

In really hot weather this desirable effect can be amplified. Instead of pushing the bed's humus-rich surface soil back into the furrow atop the seed, why not fill that furrow with pure fine compost? Compost holds several times more water than even clay will. Compost also will almost never form a crust or pose any obstacle to the shoot as it seeks the surface. Seed placed atop moderately compressed soil and covered with compost needs watering far less often, perhaps not at all.

When to water

Watering lowers the soil temperature. If the water is cold, it can take quite a few hours before the sun warms the earth back up. Water evaporating from the soil's surface also lowers its temperature. This means the worst possible time to water germinating seeds is late in the day. Doing so will leave the bed cold all night, and it will not warm up again until well into the next morning. The best time of day to water a seedbed is late morning, when the sun is getting strong and can reheat the soil as fast as possible. The next best time is midafternoon, when there is still enough time for the sun to reheat the bed before it starts setting.

Chitting

"To chit" means either to presprout seeds or to green-up seed potatoes, getting the shoots to start emerging from the eyes. Chitting brings seed partway through the sprouting process before it is put into the earth. Chitting prevents germination failures.

I mentioned earlier that, while sprouting, the seed first puts down a root to insure it has a moisture supply. Only then does it make a shoot and head for the light. Seeds usually need to be kept moist only until they are putting down their first root. The time required for this to happen depends on the species, the vigor of the particular batch of seed, soil temperature, and weather (if the seed goes into dryish soil, there might be a wait until rain initiates the sprouting process). Suppose you bring your seeds to the point where they have started making a root *before* you put them into the soil? This is chitting.

When you chit, you initiate sprouting under nearly ideal conditions. Then, after the root emerges but before it gets long (because the root is delicate and snaps off easily), you gently place the sprouting seed in its drill or hill and cover it. Unless the soil is so cold that all development completely stops, within 24 to 36 hours the root tip will have penetrated far enough that the seed will be immune to drying out.

Another advantage of chitting is that you won't sow dead or nearly dead seed and then go on hoping it'll germinate, thus losing a valuable window of opportunity. You're able to observe the initial sprouting process. With most species, if, after moistening the seeds, four or five warm days pass without any sign of a root emerging, you can be pretty sure you've got dead or extremely weak seed. If you've chitted 20 seeds and only 10 of them sprouted weakly, you know two things: although the germination conditions you've provided are not germ-lab perfect, still, 50 percent germination is pretty poor stuff. You'd best start a heap more seeds of that packet because in-the-field performance is going to be poor. Or else try other seed.

Chitting beans and peas. It's especially helpful to chit beans and peas when you're sowing early, in soil not yet quite warm enough. Presprouting indoors preserves a lot of the seed's energy to help it get past the harsh conditions outdoors.

Count out three seeds for every plant you'll ultimately want, and put them in a glass jar. Soak in tepid water for no more than six to eight hours. Attach a circular piece of plastic fly screen over the mouth of that jar with a stout rubber band. Then drain the water, rinse with tepid water, and drain immediately and thoroughly. Place the jar on its side near the kitchen sink. Room temperature should exceed 65°F (18°C). Twice a day, gently rinse with tepid water and drain. In three to four days, roots should be emerging. Plant them outside before the rapidly developing (and brittle) roots grow longer than the seeds. In each spot you'll want a plant to grow (the British call this a "station"), gently place at least two sprouting seeds about 1½ inches (four centimeters) deep, if possible with the root pointed down. Cover with loose soil. They'll be up and growing in a few days.

Chitting cucurbits. Melons and cucumbers are heat lovers; their germ temperature has to exceed 70°F (21°C) and goes best over 75°F (24°C). Squash will sprout at 65°F (18°C) but prefers at least 70°. To hold this temperature I use the same germination box I described in the section on raising transplants in Chapter 3. Some gardeners put the seeds on top of a hot water tank. An electric hot pad set on low might not get too hot for them, but don't let the temperature exceed about 85°F (29.5°C) or the germination percentage will decline. If it's warmer than 90°F (32°C), you may kill them.

Fold a section of paper towelling, roughly one foot square (30 by 30 centimeters), into quarters, dip it into tepid water, and let most of the surplus water drip out. Place four seeds for every plant you'll ultimately want into each of the folds. Place the towelling inside a small, airtight, plastic storage container or a small, sealed, plastic bag and put that in a warm place. After three days, begin checking the seeds twice a day. As soon as the roots are emerging, and before they exceed the length of the seed, plant them. Carefully place three sprouting seeds at each station or hill you'll want a plant to grow. Gently place the seeds into a small hole about 1½ inches (four centimeters) deep, scooped out of loose damp soil. Try to have the roots pointed down. Cover them with loose soil and *do not water unless the earth is bone-dry*. Unless you want the seeds to die from cold and/or powdery mildew, do not water until the sprouts emerge. You don't even want to see rain before then.

Chitting small seeds. Sowing small seed in summer's heat can be difficult as the surface soil dries out quickly. To give them a head start, place the seeds in a small jar and soak them. Do not submerge the seeds any longer than eight hours or you may suffocate them. Cover the jar with a square of plastic window screen held on with a strong rubber band. Gently rinse with tepid water and immediately drain them. Repeat this rinse and drain two or three times daily until the root tips begin to extend from the seed coats. Then immediately sow the seeds because the emerging roots will be increasingly prone to breaking off and, worse, will soon form tangled inseparable masses.

If the crop will occupy a good many row feet, presprouted seeds that have achieved only slight root development may be poured over a quart (liter) or so of composty soil and then gently tossed like a salad. Sprinkle this mixture into the bottom of a compressed furrow and cover it to the correct depth. If the weather is hot and dry, immediately water lightly and consider erecting temporary shade over the row until the seedlings appear. You will probably only need to water once or twice unless the weather turns scorching, as the seedlings will emerge in a surprisingly short time. This same technique helps get them up and going sooner in the chilly soils of early spring as well.

Getting a uniform stand

If you want many row feet of uniformly spaced carrots or radishes or lettuces, first chit the seeds on your kitchen counter. As soon as they're barely sprouting, blend the seeds into a homemade starch gelatin. Then, with a few cents' worth of jerry-rigged equipment, imitate what commercial vegetable growers call fluid drilling.

To make the gel, heat one pint (500 milliliters) of water to boiling. Dissolve into it two to three heaping tablespoons (30 to 45 milliliters) of cornstarch. Place the mixture in the refrigerator to cool. The liquid will set into a gel about as viscous as a thick soup. Don't let it get cold — only cool enough to set. (You may have to make this gel a few times until you find the correct proportions of water to cornstarch. Too thin and the mixture will not protect the seedlings from damage; too thick and you won't be able to uniformly blend in the sprouting seeds without damaging them.)

Using a spoon, gently and uniformly stir the sprouting seeds into room-temperature gel soup. Put the mixture into a sturdy, one-quart-sized (one-liter) plastic (zipper) bag and, scissors in hand, go out to the garden. After the furrows have been made, cut a small hole (under three sixteenths of an inch or four millimeters in diameter — better smaller than too big) in a lower corner of the bag. Walk quickly down the row, bent over low, uniformly dribbling a mixture of gel and seeds into the furrow, much as you'd squeeze toothpaste out of a tube. Then cover them as usual.

It may take you a few trials spreading gel without seeds to get the hole size and gel thickness right, but once you've mastered this technique, I think you'll really like it. You'll need a lot less seed per length of row than you previously did; you'll get a far higher percentage of emergence, far more quickly; and you'll save lots of time thinning.

Farmers use some costly and complicated seeding equipment to avoid wasting seed and reduce thinning. Garden magazines and seed catalogs sell various planting machines that don't work as well as their propaganda suggests. Here's a technique that does work well and that costs nothing. When sowing small seed in rows, put from a quarter to a half teaspoonful (1.25 to 2.5 milliliters) of seed into a quart or two (one or two liters) of sifted compost or fine composty soil on the dryish side and thoroughly blend the seed in. Then distribute this mixture down the furrow. You'll do much less thinning.

I know from experience that a heaped quarter teaspoon of strong carrot seed carefully mixed into three quarts (three liters) of fine compost will start two to four carrots per inch (2.5 centimeters) of furrow down about 50 lineal feet (15 meters) of row. There's no way to be exact about this, however, because the size of carrot seed and the real percentage of seedlings that will emerge varies greatly. Seed size also varies from species to species. Half a teaspoon (2.5 milliliters) of *strong* lettuce seed might be enough for twice that length of row.

Overseeding and then thinning

Compared to the environmental menaces it faces, a newly emerged seedling is a tiny, weak thing. It has just been through a gruelling, exhausting process not all that different from the birth of a baby. It has run a gauntlet of soil diseases. Plant eaters like slugs, snails, woodlice, and earwigs are waiting to take a chomp out of it. These animals are huge compared to the size of a newly emerged seedling. A few bites can remove so much leaf mass that the seedling dies.

Some new seedlings are always going to die. If you sow one seed and expect to get one plant growing from it, prepare to be disappointed; you'll soon be going to the garden center and purchasing seedlings. Transplant survival seems far more certain because once a plant has achieved a few true leaves, it is much more immune to predation. Better than transplanting, however, is to sow three or five (or sometimes even more) seeds for every plant wanted. Then when nature kills off a few, you can consider it to be helping you with your task of thinning.

Overseeding imitates nature's way. But to grow properly, each veggie needs a minimum of space; to grow magnificently, each veggie needs a lot more space than the bare minimum. Crowded radishes will not bulb at all. Crowded carrots mostly develop tops. Densely packed bush beans set small, often tough, and frequently irregular pods that take a long time to pick. Crowded tomatoes, zucchini, and cucumbers stop setting much fruit.

I wish it were possible to produce a packet of 100 percent germination seed from which every seedling becomes a perfect plant. But, alas, this is not the nature of vegetable seeds. It reminds me of the old American legend about Squanto, of the Patuxet tribe, who taught the Pilgrims how to grow corn. "Dig a hole," he told them. "Put in a dead fish and cover it. Plant four corn seeds well above the fish — one for the worm, one for the crow, one to rot, and one to grow." Squanto should also have said, "If the one don't rot or the crow don't come, there's thinnin' to be done."

I've met gardeners who cannot force themselves to thin, which to them seems a cruel act, almost like murdering children. I entreat you, you gentlest of persons, to reconsider the nature of plants. Thinning seedlings is not like drowning unwanted kittens. Vegetables don't mind being thinned. They actually like it. Thinning helps them. Your vegetables understand you must sow several seeds to get a single plant established because they do the same thing themselves, only more so. In order to get a single offspring growing to maturity, wild plants sow a hundred times more seeds than a gardener will ever sow. And wild plants thin themselves in less merciful ways than the gardener will.

Here are two examples that illustrate what I mean. Consider the propagation of any wild member of the cabbage family. Coles shoot up lanky stalks covered with enormous sprays of yellow flowers. Each flower then becomes a skinny seed pod containing half a dozen or so seeds. These seeds fall to earth and sprout after getting thoroughly soaked. Often all the seeds within a single pod sprout at once, split their pod open by the combined force of their germination, and come up as a little cluster. The seeds within each pod might result from one visit of one bee, but each seed might still be fertilized by a pollen grain from a different plant. Thus every seed in the pod may be genetically different. Cole seedlings are weak and small, but by coming up in a cluster they combine forces. A clump of seedlings may emerge where a single seed would fail to come up. Then each seedling in this group competes for water, nutrients, and light. The most vigorous one dominates the space as the rest slowly die. The winner is the one best suited to reproduce. One wild cabbage plant may have to disperse ten thousand seeds for one to survive and breed.

Now consider how the cucurbit family does it. A wild cucumber or wild melon makes quite a few fruits, each one full of seeds. After the fruit dries out, these seeds sprout in one huge cluster. Like the ones in the cabbage pod, the seeds within that single fruit are genetically different. The one seed that dominates the area is the one that grows to produce more seed next year. All the hundreds of others die after much struggle.

I hope that is sufficient argument to convince you, gentlest of readers, that thinning agrees with nature's plan. Your way of doing it is, in comparison, quick and merciful.

Generally, you should thin in three gradual steps over three to five weeks. This progressive thinning helps you avoid ending up with gaps in the row even if germination is low, if bad weather slows early growth, or if you lose seedlings to predation. Another general rule: The bigger the seed, the more certain the germination and the smaller number of extra seeds you need to sow for each plant wanted. I always sow two to three large seeds (corn, beans, squash, melons, cukes, radish) for every final plant wanted; I sow four to six small seeds for every plant wanted. When it comes to really tiny seeds like celery or basil, I might sow as many as ten seeds in a cluster at a station where one mature plant will be wanted. Except, that is, when I am sowing pricey hybrids that are sold by count, usually for so much per 1,000 seeds. Hybrid seeds are usually so vigorous (if they are fresh) and so uniform that two or three seeds are enough to start each plant desired when you are direct-seeding.

Where seeds have been sprinkled into furrows (drills), a few clumps of seedlings may need to be thinned immediately after emergence. During their first week, weaker seedlings may thin themselves out for you by falling prey to damping-off diseases and insects. Another week passes, and with most species the first true leaf should be developing. Now is the time to help thin them out so the seedlings stand about one inch (2.5 centimeters) apart. For big seeds, the initial thinning should be to at least 1½ inches (four centimeters) apart.

With open-pollinated varieties, within two weeks of emergence the more vigorous individuals will stand out. At this point, remove the weakest seedlings to give the stronger ones more unencumbered growing room. Should a remarkably vigorous plant appear, pull it out too. It is almost certainly an unintended intervariety cross-pollination, a super-hybrid. It'll outgrow everything, but what a cabbage-kale cross or a zucchini-pumpkin cross finally produces will almost certainly be disappointing.

Once the little plants are established (three true leaves and growing well), they are relatively immune to sudden loss from insect or disease and may be thinned to the desired final spacing.

Thumbprints on stations are an excellent way to start small seeds that will grow large plants such as cabbage, broccoli, Brussels sprouts, cauliflower,

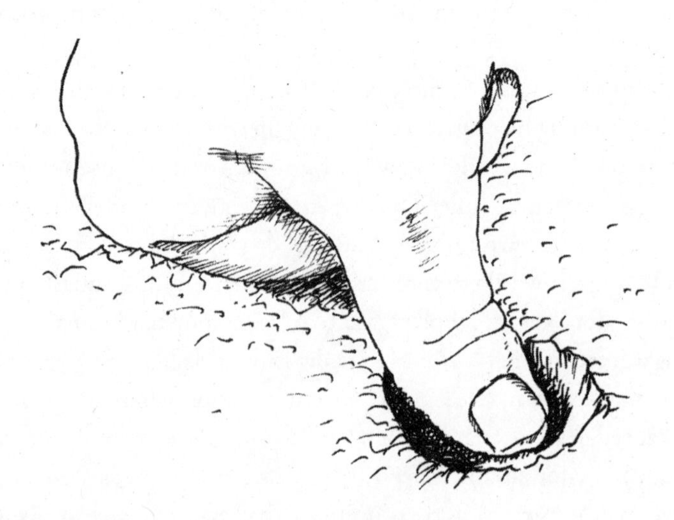

Figure 5.4: *Making a thumbprint depression for sowing seeds.*

Chinese cabbage, celery, celeriac, and kale. I will even endure the tedium of doing this for smaller-sized plants, like kohlrabi, when using expensive hybrid seed because the seed cost is high enough to make the extra time spent worthwhile. Spread complete organic fertilizer or chicken manure compost and dig the bed. Let a few days pass to allow the soil to resettle and the capillary connections to be restored. Then place manure/compost on the surface and rake it level. With your thumb, press a small indentation slightly over half an inch (1.25 centimeters) deep. Carefully count out four or five seeds into the bottom of that depression and then cover them with a bit of loose soil. Use a fingertip to trace a six-inch-diameter (15-centimeter) circle in the soil, with that group of seeds right in the center. That'll mark the spot so you won't forget where they are. Put these stations on whatever final spacing you want the area to be — 24 by 24 inches (60 by 60 centimeters), or whatever.

As soon as they've germinated, thin the clump to the best three or four, and then proceed with the same gradual thinning pattern used for other crops. When what has proved to be the best plant in the cluster has three true leaves, it should stand alone on that spot. If you are sowing tiny seeds like celery, with a naturally low germination rate, make the depression shallower and put in more seeds, as many as ten per spot. And instead of covering them with soil, use a thin mulch of fine compost or aged manure that holds more moisture.

Seedlings should never ever be allowed to strongly compete with each other for light, water, and nutrients. I can't stress this enough, gentlest of readers. When you sowed those seeds, you undertook to maintain the terms of the contractual agreement we humans made with that species long ago when it agreed to become our vegetable and we agreed to prevent it from having to compete. If you don't hold down your end of the bargain, the vegetables won't be able to do their best.

Saving on seed purchases

When you allow for seeds that don't germinate and for thinning, you can see that you'll be buying at least three or four times as many seeds as the number of plants you want to end up with each year. But you'll save money if you buy even more seeds and use them over several years. Check the prices in a mail-order catalog that offers more than one size of packets. You'll see that as the packet size increases, the unit price goes down — a lot. Here's an example using 2005 prices. Johnny's Selected Seeds sells a mini-packet containing about 120 seeds of hybrid kohlrabi for $2.85. A thousand seeds of this same variety costs $7.60. If you purchased 1,000 seeds in mini packets, you'd need eight or nine of them, and eight minis would cost you $23.75, not $7.60. Quite a difference! To risk saving this much money you need fair certainty about two things: one, you are going to grow that variety for more than one year; and two, the seed isn't going to die before you use most of it up. Notice I said "most," not all. That's because if you only used half of those 1,000 seeds, it still would work out to be cheaper to purchase 1,000 than to purchase three mini-packets containing only 360 seeds.

I grow a decent-sized kohlrabi patch every autumn, at least 100 of them. To directly seed 100 plants I use 300 seeds. Thus I am certain to use 1,000 seeds in three years. The whole trick is understanding how to preserve their germination.

Assume for the moment that the seed is vigorous, meaning its seed coat contains an abundant stock of high-quality food. The embryo sleeping within the seed will use most of that food reserve to construct a new plant, but some reserves are consumed to stay alive until the embryo does start sprouting. If the seed is stored too long, there won't be enough reserves remaining to allow the embryo to fully develop when it does finally sprout. In that case the seedling will run out of food and die before it gets established. Starvation can happen in dormancy if you wait long enough. Starvation can even happen after the shoot emerges into the light. If the plant is a "dicot," the type that starts out with a pair of small fat leaves that don't look like the next set of leaves it makes (such as bush beans, radish, lettuce, and beet), then it is not capable of producing more food than it is consuming until it has developed its first true leaf. The first true leaf is also built from food reserves. A seedling with insufficient reserves may emerge; sit there with two little leaves, stunned, exhausted; and then, unable to continue, fall over and die. It might seem damping-off or some other disease killed it, but actually it was death by starvation.

How rapidly a dormant embryo consumes its food supply, and how much or little that supply becomes degraded over time, depends on storage temperature and humidity. There's a rule of thumb about seed storage: Every 10°F (5°C) increase in temperature above what might be called "standard conditions," combined with a 1 percent increase in the moisture content of the seed, cuts the storage life of the seed in half. On the other side, for every 10°F decrease in storage temperature, combined with a 1 percent decrease in the moisture content of that seed, the storage life of the seed doubles.

Standard conditions are assumed to be 70°F (21°C) with enough humidity in the air that the seed stabilizes at 13 percent moisture by weight, which is what usually happens in a temperate climate. If we could cut the storage temperature to 60°F (15.5°C) and at the same time dry the air a bit so the moisture content in the seeds dropped to 12 percent, the seeds would last twice as long. If you keep the seeds at 80°F (26.5°C) in a more humid environment, where they'll hold 14 percent moisture, their life span will be cut in half.

If you keep your seeds in a cardboard box in a steamy greenhouse, they won't last a year. If you keep them on a closet shelf in an unheated bedroom, they will probably last longer than they would at standard conditions. When I ran Territorial Seeds, I knew Oregon's climate was extremely humid in winter and rather hot in summer. I built a climate-controlled storeroom that held the seeds at 50°F (10°C) with 50 percent relative humidity. At RH 50 percent, the seeds dried down to about 10 percent moisture, so my seeds lasted at least four times longer than if they had been stored under standard conditions. Johnny's and Park Seed Company do this, too.

One other thing shortens seed storage life: the stress of change. Seeds will do much better if their temperature and moisture content don't alter.

Here's how you can keep seeds at home under nearly ideal storage conditions. First, go to a hobby/craft shop or chemical supply house and buy a pound (or a half kilogram) of silica gel desiccant crystals. These don't cost much. The crystals are usually dark blue when they are dry; the color slowly changes to light blue and then finally to pink after they have soaked up all the moisture they can hold. Silica gel can be reactivated many times as long as it is not heated beyond 230°F (110°C). If the crystals aren't dark blue, put them into a baking pan, pop this into the oven set over 212°F (100°C) but under 230°F, and bake them for a few hours until nicely blue. As soon as they have cooled a bit, put them into something airtight so they don't pick up moisture.

Now get some large airtight containers to hold your valuable seeds. I have used one-gallon (four-liter) restaurant mayonnaise jars and half-gallon (two-liter) wide-mouth canning jars. I currently use a large, rectangular, two-gallon (eight-liter), plastic storage box that seals nearly airtight. For each half gallon of seed-storage volume, put in about one cup (250 milliliters) of silica gel. Keep the crystals in a cloth sachet or a small, clear, open-topped container so that you can conveniently monitor their color. This addresses the low-humidity part of the storage. If the gel turns pink, take it out and reactivate it; after cooling it off, return it to the seed container.

The next thing to address is the cool and stable part of the equation. I keep my big plastic tub in the refrigerator. I confess I had to squabble a bit with Muriel — more than once, in fact — to claim half a shelf, maybe one eighth of the entire space inside *her* refrigerator. The root cellar would also be a good place for this, if you have one. Least ideal, if your partner won't cooperate and share the fridge, is an unheated room during winter and the coolest place you can find in summer, such as under the house.

There is always more uncertainty about the storage life of seeds you purchased than of those you grew yourself. The seed you buy is new to you, but for how many years, and under what storage conditions, has it been sitting on some seed-warehouse shelf? How much of its potential storage life is left? And how good was that seed from the start? Some years the weather conditions don't cooperate and the seed fails to fully mature, or it may be rained on a few times while it is drying out, or the field that was used to grow it might not have been as fertile as it should have been. When you grow your own seed, you know a lot more about it.

Growing your own

This is an era of huge controversy about the seed industry's consolidation, about genetically modified (GM) seeds, about hybrid seeds and heirloom seeds, terminator technology, and patenting, with lawsuits claiming huge amounts in damages from helpless farmers who merely grew a bit of seed. At bottom the issues are about the ability of our system of agriculture to continue if we lose our genetic materials, and perhaps about our enslavement to corporate interests.

All this ruckus provides a business opportunity for politically correct mail-order seed suppliers posing prominently on the side of the angels. Their catalogs suggest that buying heirloom seeds allows gardeners to grow their own. But this claim is not entirely true. With some vegetable species, saving seed is so simple to do and takes so little effort that I've wondered why so many people don't do it. For other species, growing effective seed is a daunting prospect best left to the professional. Easy or daunting is mostly determined by how the species fertilizes (pollinates) itself. Some species self-pollinate; others exchange pollen with neighboring plants (see Figure 5.5).

Self-pollinated species inbreed generation after generation without any harmful consequences. These species are usually quite stable; from generation to generation there is little or no change. To save seeds of these species, you simply save the seeds. A few kinds of self-pollinated vegetables (such as peas and lettuce) have a slight tendency to outcross and should be isolated from other varieties, but 20 feet (six meters) is usually enough separation. All these are fine candidates for saving your own. Seed can be saved from a single plant, and even from a few pods or a single fruit on one good-looking plant, and this can be done year after year, with no problems.

Other species must exchange pollen (outcross) or they become so weak (inbred) they can no longer grow or reproduce. It takes more skill and considerably

Vegetables by method of pollination

Self-pollinating vegetables (Inbreeding)	**Pollen-exchanging vegetables (Outcrossing)**
Beans	Alliums (onions, leeks)
Eggplant	Asparagus
Endive/escarole/chicory	Beet (beetroot) and chard (silverbeet)
Garlic (no seed)	Brassicas, refined (broccoli, Brussels sprouts, cauliflower, turnips)
Lettuce	Brassicas, unrefined (kale and kohlrabi)
Peas	Chinese cabbage
Pepper	Corn
Tomato	Cucurbits (melons, pumpkins, squash, cucumber)
	Mustards
	Okra
	Rutabagas (Swede)
	Spinach

Figure 5.5

more land to grow seed for species that must outbreed. First of all, seed raisers must wisely select which plants they will allow to contribute to making seed and which ones they will take out. Selection is necessary to prevent the variety from showing too many unwanted characteristics. But selection must not hold the gene pool too rigorously to an exact form because there's another pitfall when growing seed from outbreeding species. To avoid what is termed "inbreeding depression of vigor," the variety needs to have a diverse gene pool composed of plants that are somewhat different from each other. If the seed-making population is either too few in number and/or too uniform in genetics, this line will die out. After only a few generations of inbreeding, a variety can become so weak that its own seeds will barely sprout and will grow so weakly that the variety becomes nonproductive. For a few extremely vigorous outcrossing vegetables that aren't highly refined, like kale or rutabagas (swedes), the minimum number of individuals needed to avoid inbreeding depression of vigor might be as few as a dozen. For corn, the minimum number of individuals needed to maintain a reasonable level of vigor may be as low as 50. For other species, 200 plants might be the minimum.

Imagine trying to maintain a full-sized cabbage variety when the minimum sustainable number in the gene pool is 200 plants. To allow for weeding out off-types, you'd have to grow a field of 300 to 400 cabbages and select the best 200 for making seed. Do the math: each plant needs a growing space of 4 square feet (0.4 square meters) per plant times 400 plants, which is 1,600 square feet (150 square meters), or 16 beds of 100 square feet (10 square meters) each. Why, that's two thirds of my active garden beds! And that many plants might make 50 pounds (20 kilograms) of seed. Clearly, growing cabbage seed is something for a market gardener to do as a moneymaking sideline and not something for the home gardener. However, there are some small-sized outcrossing varieties — scallions (spring onions), for example, or kohlrabi or bulbing onions — so for these vegetables a fair number of individuals can be grown for seed on a garden scale.

There is a money-saving strategy for limited home-garden seed production of outcrossing species. Instead of trying to grow seed for a variety indefinitely, start with the highest-quality seed you can buy and then grow your own seed from it, but only for one generation. The seed yield from only a half-dozen plants would supply you (and your neighbors) for a decade. The variety wouldn't become severely inbred in only one generation, and if some of that seed production were kept under excellent storage conditions, it might last more than a decade, depending, of course, on the species and how vigorous the seed was to begin with. When that lot was used up or lost its ability to germinate strongly, you'd buy another packet of commercial seed and do it all over again.

In Chapter 9 I'll suggest seed-raising methods for each vegetable. I also recommend two books if you want to grow your own seeds: Susan Ashworth's *Seed To Seed* and Carol Deppe's *Breed Your Own Vegetable Varieties*. Both are listed in the Bibliography.

Dry and wet seed

There are another two basic vegetable-seed divisions that have nothing to do with their method of pollinating: "wet seed" and "dry seed." Dry seed forms in pods, in clusters on the stalk, or in the dried flower structures — examples are beans, peas, lettuce, mustard, spinach, beet, okra, etc. Wet seed forms in juicy fruit that is still full of moisture when the seed has matured. These include squash, pumpkin, cucumber, melon, tomato, pepper, etc.

The key to getting vigorous dry seed is to let it mature fully but to keep it drying down steadily while it matures. Should the ripening process continue over weeks, sometimes the only way to keep drying seed from being remoistened is to yank the entire plant, shake the soil from the roots, and move it under cover, spreading it out on a tarp or hanging it above one to catch shattering seeds. If the species makes large seed in big pods, such as peas or beans, the gardener can produce seed of a quality no commercial grower could dream of; we can pick each pod at the point in maturation when the stem end of the pod withers and it is clear that the plant's sap is no longer flowing into that pod. The seed won't have dried down hard at that stage, but if any further nutrition is added to the seed's food reserve, it'll be coming from material held in the pod itself, translocated into the seeds as the pod withers completely. When I am growing bean and pea seed, every day I put a few pods that have dried to the right point into my pocket and take them up to the house to finish drying in a large bowl.

One common mistake made by gardeners growing dry seed is to accelerate drying down by putting the pods or whole plants into an unusually warm place. Remember the data I provided earlier in this chapter about seed life? If the seed is not dry, then it is moist. At high moisture contents and at high temperatures, the seed ages rapidly. It's best to dry it down on the cool side.

The key to getting vigorous wet seed is to let the fruit become fully ripe on the plant before harvesting it — allow these dead-ripe fruits to continue ripening nearly to the point of rotting on the vine before you extract the seeds. If it is tomato seed, take your seeds from overripe fruit that hid unnoticed, or let a bowl of soft ripe fruit sit on the counter for a few more days before extracting the seed. If it is a melon, make sure the fruit chosen for seedmaking honestly slips the vine. If it's squash, allow the fully ripe fruit to cure in the house for a month or so before extracting the seed.

Seed from hybrids

There is a lot of anti-hybrid seed propaganda in circulation. Much of it is misleading, put forth by people who want to profitably sell you low-quality open-pollinated seeds; some anti-hybrid information is true. I know the anti-hybrid story well. I once ignorantly believed it. But after getting into the seed business, I gradually observed the truth of the matter, as demonstrated on my own trials ground. I also learned a few things from my own breeding work that started with hybrid varieties.

Anti-hybrid propaganda says that gardeners can't grow their own seed from hybrids. This is a half-truth on several levels.

First of all, hybrid seeds will not replace OP varieties in some species. Hybrids rarely can be mass-produced for species that self-pollinate. The flowers of this type of plant are structured in such a way that they pollinate themselves before opening enough to allow any access by insects. The only way to produce a hybrid variety in one of these species is to perform microsurgery on the unopened flower before it pollinates itself. This is done on hands and knees in the field, aided by a powerful magnifying glass. Each delicate operation will, if successful, result in a few seeds. The making of each seed costs several dollars worth of the breeder's time. This sort of hybridization is done only to breed new OP varieties. Thus there are no hybrid seeds being sold for beans, peas, lettuce, etc.

But the home seed savers usually concentrate their efforts on just these self-pollinating species because large plant populations aren't needed to carry the line forward indefinitely.

When it comes to the solanums — tomato, pepper, eggplant — which are also self-pollinating species, there are hybrid varieties that are made surgically, painstakingly, by hand. However, when someone makes hybrid tomato seeds, each pollination results in a hundred seeds or more forming in one big fruit, not six seeds inside a tiny lettuce flower or bean pod. In commercial production, each of the jillions of flowers on a tomato or pepper or eggplant has to be worked over every single morning because each flower must be hybridized before it self-pollinates. As this is done, a little flag has to be tied to the stem of each flower cluster. Hybrid solanums can be produced only where labor costs are extraordinarily low. In the 1980s this was Guatemala. These days they're being made in China. Even with minuscule labor rates, hybrid tomato seed still wholesales for close to a thousand dollars a pound (450 grams). Hybrid peppers are only a bit cheaper. There are minor advantages to be gained by growing hybrid solanums; they may be a bit more productive and can resist certain diseases better than any known OP variety. Hybrid peppers and eggplants are actually quite useful, maybe essential, in maritime climates where the conditions are so marginal for the species that OP varieties are not productive. Elsewhere, most gardeners have no need of them. The existence of hybrids within these species are not threatening the continued availability of OP varieties.

In some species that naturally outcross, breeders have worked out clever methods of mass-producing hybrid seeds. Commercial growers prefer these varieties because they are more vigorous and perfectly uniform. Despite this kind of seed's high price, higher yields make the grower more money. Because the economic superiority of hybrids has caused interest in using OP varieties to virtually disappear within the commercial trade, the remaining OP varieties are produced only for the noncritical home-garden market — meaning we gardeners are usually offered degenerated, ragged, low-quality, poorly bred material. For example, there are very few decent OP brassicas offered, although there are a few left. Johnny's Yellow Crookneck and Costata de Romanesco are the only decent OP summer squash left that I know of.

Anti-hybridists accuse seed companies of intentionally making hybrids that fail to produce any seed, trapping the customer into buying more seed the next year. Is that true? Like most such issues, the answer is "yes, and no." It is normal practice for amateur plant breeders to use commercial hybrids as starting platforms to breed new OP varieties. My friend Tim Peters bred Umpqua, an OP broccoli, from Green Valiant, a hybrid. Tim told me that in the first generation, Green Valiant did not make many seeds — only a few hundred in total on half a dozen plants. That's because mass hybridization is often done by breeding self-sterility into a line so it just about can't self-pollinate. It becomes what is called a "self-sterile parent line." This parent line is then grown in a field with another variety that is genetically highly uniform; the two varieties are definitely not mutually sterile. Enough rows of the second variety are grown in the field so the bees can fully pollinate the self-sterile rows. The self-sterile variety then sets heaps of seeds, nearly all of which (except for the odd one here and there) are formed by crossing with the second variety. The seed from the self-sterile parent line is then a perfectly uniform, highly vigorous hybrid, with virtually every seed being a cross between two perfectly uniform inbred varieties of little vigor.

Tim said that after that scant 100 seeds, the next generation set seed normally and abundantly. From there it took but a few generations and Tim had a uniform OP variety that made broccoli quite similar to Green Valiant, albeit not quite as uniform or vigorous. But that lack of uniformity and vigor is the basic nature of OP varieties, and why, when offered a hybrid against an OP variety, the commercial grower will almost inevitably opt to pay a higher price for hybrid seeds.

I had a similar experience breeding carrots. I was interested in making an OP out of a hybrid carrot with unusual and useful traits, so I planted some overwintered hybrid carrot roots in spring and let them make seed. But the flowers that formed produced not a single viable seed. Was the anti-hybrid propaganda about to be proved correct? At the suggestion of Carol Deppe, whose book is listed in the Bibliography, the next year I again planted out these hybrid carrots, but alongside this row was another row of an OP variety with a similar appearance. The result: both rows set heaps of good seed. The seed from the hybrid carrots could only be a cross between it and the OP. So from that seed I started breeding, and in a few generations I had a quite uniform carrot variety that was everything I had hoped to make.

Other amateur breeders have started out with hybrid variety conversions by allowing two similar hybrids to make seed in proximity, resulting in lots of seed. I have made OP seed from a hybrid sweet corn called Miracle; it produces a remarkably uniform stand nearly as productive as Miracle in the first generation. Had I been a corn breeder, I could have developed a dozen rather similar new OP sorts from Miracle.

Sometimes I suspect that expensive hybrid seed is not really hybrid seed at all. Oh, it is labelled as hybrid, and sold for hybrid prices, but judging by how the seed grown from it behaves, the variety may as well be OP. When I produced seed from 100 plants of Kolibri F1 purple kohlrabi, the patch made heaps of seed and that seed produced kohlrabi as uniform, as productive, and as tasty as Kolibri F1 had been. I have heard similar stories about so-called hybrid tomato varieties. Keep in mind, when considering the possibility of falling to ethical temptation, that hybrid tomato seeds might be 20 times more expensive than OP varieties.

However, as I mentioned earlier, the average gardener is not going to grow seeds for varieties that are pollinated by wind or insect. When seed savers do try it on a home-garden scale, what almost always happens is that they begin working from too small a gene pool, using too few plants. In subsequent generations they proceed to make mistakes about which plants to take out and which to permit to remain. The result is nearly certain — the variety is ruined within two generations. For that reason I am not willing to waste time and valuable garden space by starting with seeds for outcrossing species produced by a "network of collaborating seed growers." That's not to say these heirloom varieties are consistently bad. Occasionally there will be good stuff. But usually there isn't. However, if the varieties are in self-pollinating species, you can be assured of success with such amateur seed.

I don't want to imply that the average person lacks the skill or intelligence to grow seed for insect- or wind-pollinated vegetable species. A century or more ago, the average person grew up on a farm and was involved constantly with the breeding of animals and plants. In that era, the home garden was a lot bigger than it usually is today. It was not awkward to grow a fair-sized patch of cabbage, corn, or carrots to make seed. Still, even in that time, most people saved seeds only from their self-pollinating species or the cruder, less refined, less carefully bred of those that outcrossed; for refined outcrossing varieties they purchased seeds grown by specialists.

Finally, the anti-hybrid propaganda says that hybrid varieties are not nutritious and don't taste as good as the old-fashioned OP sorts. This can be true, but isn't the whole truth. What is also true is that the modern OP varieties, those bred after, say, 1870, are not as nutritious or as tasty as the old-fashioned heirlooms were. For this insight I wish to thank Dr. Alan Kapuler, an American biophysicist who in the 1960s dropped out of academia, homesteaded in Oregon, and later was instrumental in the creation of Seeds of Change. One night over conversation, Alan suggested the reason for the decline in nutrition and taste to me, using the following example.

For thousands of years, each family raised seed for its own unique varieties, selected over generations to suit the needs and tastes of the family members and the soil they grew on. The nutritional quality of each variety was different. When two varieties are grown on highly fertile soil, one will probably be more nutritious than the other. In those days before antibiotics and mass sanitation, children had to run a gauntlet of childhood diseases. A great many of them died before reaching breeding age. Some families lost most or all of their children before they reached the age of seven. Others rarely lost any kids to disease. The family with better vegetable (and cereal) varieties tended to have children who survived to carry on those varieties. Less-nutritious varieties tended to disappear.

Thus, over thousands of years, varieties used for human food were selected for producing good health in those who ate them. One indicator of possessing a high level of nutrition is good taste; humans naturally select for flavor.

Then, by about 1870, Europe and the United States had gone through the industrial revolution. Farming ceased to be a subsistence activity and became a business. Before 1870, more than 90 percent of people in North America lived on self-sufficient farms or in small villages where the houses all had half acre or acre blocks and big veggie gardens. After 1870, increasing numbers of farm families moved to the city and got jobs. The farmer now grew marketable crops and purchased most of the family's food. Many farms no longer had a kitchen garden of any size.

Before 1870 there were many thousands of varieties, maybe many tens of thousands if you count the ones saved by only one family or used in one small district. For example, I have an 1863 book describing American vegetable varieties; it contains 28 kinds (not merely similar varieties of the same kind) of radishes. After 1870, vegetable variety qualities were selected and bred to become ever more suited to profit-making, and there were ever fewer of them.

So yes, it is true on average that modern hybrids are not the same nutritious, fine-flavored vegetables that the old-fashioned OP ones were. But it is also true that the OP varieties developed since 1870 are not either. And, alas, it is true that much of what passes for "heirloom" OP varieties these days is really just the commercial varieties used before World War II or relatively modern commercial varieties that have been misnamed by unethical primary growers producing cheap garden seed.

I regret that this is an uncertain world and sometimes our choices are difficult, between "bad" and "worse."

CHAPTER 6

Watering ... and not

Before the 1930s, few farms had electricity. Many vegetable gardens were grown without running water. Before 1880, when 90 percent of all North Americans (and probably Australians and New Zealanders) lived on family farms or in tiny villages, it would be a fair guess that over 90 percent of all vegetable gardens were grown without running water. In those days, after a few weeks without rain, many would begin talking about drought and of how their gardens or crops were suffering. But other gardeners in the vicinity wouldn't be complaining much.

In 1911 John Widstoe wrote *Dry Farming*, a book about large-scale farming in semi-arid places. Widstoe had a different take on what a drought (or "drouth," as he called it) actually is. He said:

> Drouth is said to be the archenemy of the dry farmer, but few agree upon its meaning. For the purposes of this volume, drouth may be defined as a condition under which crops fail to mature because of an insufficient supply of water. Providence has generally been charged with causing drouths, but under the above definition, man is usually the cause. Occasionally, relatively dry years occur, but they are seldom dry enough to cause crop failures if proper methods of farming have been practiced. There are four chief causes of drouth: (1) Improper or careless preparation of the soil; (2) failure to store the natural precipitation in the soil; (3) failure to apply proper cultural

> methods for keeping the moisture in the soil until needed by plants, and (4) *sowing too much seed for the available soil-moisture.*[emphasis added]

I emphasize Widstoe's fourth point because this is the factor that most pertains to veggie gardening. I also emphasize it because choosing plant spacing is the single most important decision the gardener will make.

These days, having piped water is normal. Veggie gardening styles have adapted to this situation and to the needs of suburbanites living on smaller lots, who buy most of their produce instead of growing it. This shift in vegetable-growing methods was named "intensive." The idea: put the vegetables much closer together in massed plantings on raised beds so the yield from postage-stamp-sized gardens becomes greater per square foot. The gardener loosens the bed to two feet deep (60 centimeters), supposedly so the root systems can go down instead of out. The high plant density sucks the soil dry so rapidly the gardener must water almost daily during the growing season. However, the yield from the amount of water used is supposed to be greater. And, of course, the gardener must make the soil super-fertile to support this intense growing activity.

The intensivists say that putting vegetable rows far apart is a waste of garden space and that gardeners do it only in foolish imitation of farmers, who have to do it so that machinery can work the field. This assertion is not correct. The reason people traditionally spread out their plants (and why farm machinery was designed to match this practice) was so that the vegetables could go through rainless weeks without damage or moisture stress. Most of the intensivist claims listed in the preceding paragraph are also not quite true, as I will soon demonstrate.

I expect many North American cities and towns will start to experience more severe water shortages, if only due to increases in population. This is also happening in the U.K., Australia, and New Zealand. But under economic stress, many people may again wish to grow a substantial amount of their own food. What little irrigation their gardens will get may have to be recycled household water or rainwater off the roof, trapped in tanks, which is how most rural Australians still live. Plenty of piped, chlorinated water may be available in some areas, but unless alternative and large-scale energy sources are developed, the cost of this water will increase with the inevitable increase in the price of oil.

Going back to old-style gardening won't actually be a sacrifice. In fact, it may work out to be a big gain for most people. Instead of having to water the garden constantly and finding that a veggie garden makes a problem about going on a week's holiday during summer, people will discover the old-style garden can look after itself. In my experience, the supposed advantages of intensive raised beds are largely an illusion. Instead of growing many small, crowded plants that take a longer time to harvest (and clean), people will spend less time harvesting larger, more attractive-looking, more delicious vegetables. Species that produce an ongoing harvest, like tomatoes and cucumbers, will surprise the intensivist by yielding a lot more toward the end of the season because when these plants are crowded together, they produce well for a few short weeks and then virtually stop yielding, their root development stopped by overcompetition.

Four spacing systems

Please study Figure 6.1, a regrettably complex table of plant spacing possibilities. The crop spacings listed in Column 1 are typically what intensivists recommend. A crop canopy forms quickly at those spacings, and from then on watering will be necessary almost daily during sunny weather. I do not recommend using these spacings. Column 2 is my recommendation when you have abundant irrigation. Use Column 3 where there is sufficient rain during the growing season to comfortably support a garden; generally this means areas that were once covered with large trees, which describes all of North America east of the 98th meridian (a north-south line running through Dallas, Texas), the densely settled east coast of Australia, and most of the U.K. except the dryish east of England. Column 4 may be used where there is a fair depth of moisture-retentive soil (over four feet/120 centimeters) and where winter rains or snowmelt have charged that soil with water to its full depth. Column 4 assumes that what little rainfall there may be during the growing season will happen with many weeks of separation between. This describes the semi-arid parts of the North American prairie states after there has been good snowmelt from winter and also describes Maritime climates with their winter rains and dry summers.

Plant spacing possibilities (in inches)‡

	Column 1	Column 2	Column 3	Column 4
Crop	**Intensive raised beds per Jeavons**	**Semi-intensive raised beds per Solomon**	**Extensive; good rainfall; raised beds *: raised rows †; on the flat ††**	**Extensive; little rain or fertigation; everything on the flat ††**
Asparagus	12 x 12	18 x 48	18 x 48–60††	18 x 60-72
Beans, fava (overwintered)	?	8 x 24	12 x 36†	12 x 48††
Beans, fava (spring sown)	?	8 x 24	10 x 36†	10 x 48
Beans, lima (bush)	6 x 6	4 x 24	3–4 x 30*	6 x 48
Beans, lima (pole)	6 x 6	36 x 36	36 x 36†	48 x 48
Beans, snap (bush)	6 x 6	12 x 18	12 x 24*	18 x 48
Beans, snap (pole)	8 x 8	12 x 24	18 x 36†	48 x 48
Beets	4 x 4	4 x 18	4 x 24*	8 x 48
Broccoli	15 x 15	24 x 48	36 x 36†	N/A
Brussels sprouts	18 x 18	24 x 30	30 x 36†	N/A
Cabbage, early	12 x 12	16 x 24	18 x 24*	24 x 36
Cabbage, late	18 x 18	24 x 30	24 x 36†	36 x 48
Cabbage, Chinese	10 x 10	18 x 24	24 x 30*	N/A
Carrots	3 x 3	1–2 x 18	2–3 x 18*	4–6 x 36–48
Cauliflower	15 x 15	24 x 24	24 x 36†	N/A
Celery	6 x 6	18 x 24	24 x 30*	N/A
Celeriac	?	16 x 24	18 x 30*	N/A
Cucumber	12 x 12	36 x 48	48 x 36†	48 x 48–60
Chard	8 x 8	12 x 24	18 x 36†	24 x 48
Collards	8 x 8	24 x 24	18 x 36†	36 x 48
Endive	?	12–18 x 18	18 x 24*	24 x 36
Eggplant	18 x 18	36 x 24	24 x 36†	30 x 48
Garlic	?	6 x 24	6 x 24*	8 x 24
Kale	15 x 15	24 x 24	24 x 36†	30 x 48
Kohlrabi	4 x 4	6 x 18	6 x 24*	8 x 36
Leeks	6 x 6	6 x 24	6 x 24*	N/A
Lettuce, heading	12 x 12	12 x 18	14 x 24*	N/A
Lettuce, looseleaf	8–9 x 8–9	10 x 18	12 x 18*	N/A
Melons	15 x 15	48 x 48	48 x 72††	48 x 72
Mustard	6 x 6	12 x 18	12 x 24*	12 x 24
Okra	12 x 12	24 x 24	18 x 36†	18 x 48 ☞

Figure 6.1

	Column 1	Column 2	Column 3	Column 4
Crop	**Intensive raised beds per Jeavons**	**Semi-intensive raised beds per Solomon**	**Extensive; good rainfall; raised beds *: raised rows †; on the flat ††**	**Extensive; little rain or fertigation; everything on the flat ††**
Onion, bulbing	4 x 4	3–4 x 18	3–4 x 24*	N/A
Onion, scallion (spring)	½ x 18½ x 18	1 x 18*	N/A	
Parsley	5 x 5	6 x 18	6 x 24*	10 x 30
Parsnips	4 x 4	3 x 18	3–4 x 24*	4–6 x 36–48
Peas, bush	3 x 3	2 x 18	3 x 24*	3 x 24
Peas, southern (bush)	3 x 20	3 x 24	3 x 36†	N/A
Peas, southern (vining)	6 x 36	6 x 24	6 x 36†	N/A
Peppers	12 x 12	24 x 24	24 x 36†	36 x 48
Potatoes, Irish***	9 x 9	8–12 x 48	8–12 x 36††	14–18 x 48
Pumpkins	18 x 30	N/A	60 x 72††	72 x 96
Radishes, salad	2 x 2	1½ x 12	1½ x 12*	N/A
Radishes, winter	Not specified	4 x 18	4 x 24*	Don't know
Rhubarb	24 x 24	24 x 48	36 x 36†	48 x 48
Rutabaga	6 x 6	8 x 24	8 x 36†	12 x 36–48
Salsify	3 x 3	3 x 18–24	3 x 36†	4 x 36–48
Spinach	6 x 6	3 x 18	3 x 18–24*	4 x 24
Squash, summer	18 x 18	48 x 48	48 x 48††	60 x 60–72
Squash, winter	18 x 18	N/A	60–72 x 96††	72 x 96
Sweet corn	15 x 15	N/A	10–12 x 30–36††	48 x 48**
Sweet potatoes	9 x 9	15 x 48	15 x 48†	18 x 48
Tomatoes, determinate	18 x 18	36 x 48	36 x 48††	N/A
Tomatoes, indeterminate	24 x 24	48 x 48	48 x 60††	60 x 72
Turnips	4 x 4	3 x 18	3–4 x 24*	4–6 x 36

* On raised beds. **Grow in hills, two to three plants per hill. *** Potatoes are hilled up as they grow.

† In raised rows. †† Most items in this group are better grown in hills; see Chapter 9 for details.

‡ To convert to centimeters, multiply inches by 2.5.

A note on spacing: Generally the first number is the in-the-row spacing; the second number is the between-row spacing. A 24 x 24 spacing on a four-foot-wide (120-centimeter) raised bed means a double parallel row down the bed the long way. A 36 x 48 spacing means a single row of plants three feet (90 centimeters) apart down the center of a four-foot-wide raised bed. ■

Comments on Column 1

Most current garden books and government agricultural publications, having been written by academics who are trained to regard earlier similar publications as "authoritative," parrot Everybody Else's recommendation to use these plant spacings. I do not recommend using Column 1. It is here only so that you may compare the opinion of Everybody Else to something I consider practical. For example, when Column 1 says "Beets, 4 x 4" it means 16 square inches (roughly 100 square centimeters) per plant. Compare that to Column 2, which recommends 4 x 18 or 72 square inches (465 square centimeters), which allows each plant nearly five times as much room.

Once ultra-crowded plants have formed a crop canopy (in other words, the crop has fully covered the bed with growing leaves), moisture loss is so rapid during sunny weather that to prevent moisture stress, you will need to give the beds at least half an inch (1.25 centimeters) of water every other day in hot spells. If the soil doesn't retain much moisture, you'll apply less water, but you'll apply it every day. I've seen sandy intensive beds need irrigation twice daily in really hot weather. This requirement tightly shackles gardeners to their gardens.

Comments on Column 2

These spacings are how I currently garden. Most of my garden, except for the sprawling vines and a block of sweet corn, is grown on 4 x 25 foot (1.25 by 7.5 meters) raised beds; the spacings reflect layouts on beds like this. This more generous spacing will still result in a crop canopy that completely covers the bed, but the canopy does not form so quickly. Because each plant has far more soil room to fill with its roots, such a canopied crop needs watering only every four to seven days in sunny warm weather. One reason this garden will go longer between irrigations is that the plants are sufficiently spaced to benefit from capillarity flow, which partly recharges their root zone with subsoil moisture.

The volume of harvest on this spacing is about 90 percent what you'd get using Column 1, but you will put in less than half the number of hours to get it. The smaller vegetables are arranged in four-foot-long (120-centimeter) rows running across the beds the narrow way. Tightest of all are salad radishes, given only 12 inches (30 centimeters) between rows, which is still far enough apart to pass a hoe between the rows until one week before harvest. Using

Column 2, the vegetables become larger and more succulent and tender because growth has not been slowed by overcompetition; consequently harvesting (and washing/trimming) takes much less time.

Comments on Column 3

In this system, small plants are grown on wide raised beds, larger ones in slightly raised single rows, sprawling ones in hills rather far apart, and crops like corn or okra, large plants that don't need to start in a finely raked seedbed, are grown on flat ground. Even after crop growth makes it difficult to walk between the rows or hills, a garden using Column 3 may go through rainless weeks without significant moisture stress. The plants have spread sufficient roots to greatly benefit from capillarity. Their ever-expanding root systems continue to discover large amounts of soil water held in storage, like money in a bank account (think of deep subsoil moisture rising through capillarity as being like substantial interest). Yields per square foot are only slightly lower than those you'll gain by using Column 2 spacing. In dry years, irrigation would increase yields, but in most years, rainfall will be adequate; watering will not be essential. Because the rows are so far apart, hoeing weeds is easy. I grew my variety trials grounds in this style and during the main growing season, when the weeds were growing their fastest, could keep a half acre weed-free with only one short morning's hoeing each week.

Column 3 has been used by English-speaking North Americans ever since the Native Americans taught them how to garden here, based on the millennia of their own experience.

Comments on Column 4

Column 4 contains my spacing suggestions for regions with relatively dry summers and with not much water for irrigation. Unirrigated gardening in such a situation works only if the soil goes into the growing season fully saturated to considerable depth, if weed competition is meticulously eliminated, and if the soil is kept loose at the surface with regular cultivation, forming what is termed a dust mulch. An inch (2.5 centimeters) of loose surface soil acts much like a straw mulch, virtually eliminating moisture loss from sun shining on bare ground. Assuming there are no weeds, the only moisture lost will be what is transpired by growing crops.

In the North American summer, rainfall often comes as intense showers and thunderstorms. If there is a crust or hard layer on the surface, much of this rain may run off. If the surface is loose, the rain will penetrate. It is essential, therefore, that a day or so after such a rain you go out with a hoe and loosen up the surface of every accessible part of the garden. This prepares the soil to fully accept the next chance shower and also helps keep the moisture now in the earth from evaporating.

When you are gardening this way, it is essential, it is vital, it is incredibly important, it is crucial (can I express it any more strongly?) that you make sure the surface inch (2.5 centimeters) or so of soil contains a fair amount of compost or decomposed manure. This will prevent crust formation and help your soil form large stable crumbs that the wind won't blow away when you make a dust mulch.

Not suffering drought

Widstoe's fourth cause of drought, "sowing too much seed for the available soil-moisture," is a little like Will Roger's comment about speculating in the stock markets: "If you want to make money in stocks, buy one that will go up. If it didn't go up, don't buy it." The reason people fail to match crop density with the available soil moisture is greed, the temptation to get a somewhat bigger harvest by having more plants in the same area. We humans tend to forget pain and loss, to hope that some recent years of higher rainfall represent what can always be expected. If, on the other hand, gardeners were to act more like pessimists and expect that every year would be a dry one, then they would plant things a bit more spaciously, more along the lines of what is suggested in Column 4. The results wouldn't be as bad as you might think, because getting more moisture in a wet summer and having more room to grow the smaller number of plants will produce larger plants and yield more per plant. True, it won't be as much as you might have had by increasing the plant population, but you'll get it more securely.

If I were gardening in the eastern half of the United States or Canada and had started the season with a traditional garden design, based on the spacings in Column 3, and if it began to look like a mighty hot, droughty summer, there are some steps I could take to greatly reduce the damage. The first thing would be a brave action indeed: I'd go out with a sharp hoe and chop every second plant in every row and harvest every second carrot or beet, no matter how small it was at the moment. This would instantly reduce moisture consumption while providing double the capillary moisture to the remaining plants. The end result might not be the loss of half the crop. By preventing severe moisture stress, I might end up harvesting a lot more. And if it should start raining, you'd be surprised at how big the plants will grow in these roomy spacings.

If it didn't start raining, I'd thoroughly wield my weeding hoe, as described above, at least weekly.

The next thing I'd do, if at all possible, would be to start fertigation, which is something that a person might want to do in any case, drought year or not, because it has such benefits. Crop scientists of Widstoe's era noticed that it took twice as much water to make a given weight of dry plant material on soils that were infertile as it did on soil that was highly fertile. And William Albrecht explained that much of what appears to be drought damage is actually nutrient deficiency induced by dry surface soil when plant nutrients are mainly located in the plowed surface layer. This can be especially true in new veggie gardens because it can take a few years for the fertility you're putting into the topsoil to work its way down into the subsoil. When plants are not finding water in the surface layers, they must feed in the subsoil, but if that is infertile ....

To more rapidly remedy subsoil infertility when you are starting a new garden in humid regions, you may want to spread a full dose of fertility-building materials at the end of summer the year before the land will become a veggie garden. This will allow the autumn rains and the spring snowmelt to carry this nutrition into the subsoil.

Fertigation

The fertigation bucket economically and effectively places moisture and nutrition below a growing plant. In the early 1990s I did dry-gardening experiments in Oregon, where the entire hot summer is dependably rainless. Each unfertigated winter squash vine grown on the spacings suggested in Column 4 yielded 20 pounds (nine kilograms). But other plants of the same variety on the same spacing that were given five gallons (20 liters) of fertigation four separate times between the solstice and the equinox (20 gallons total) yielded 50 pounds each (22 kilograms). Thirty pounds (13 kilograms) of additional squash for 20 gallons of water seems a pretty good exchange to me.

Fertigation is especially useful for big plants: tomatoes, melons, cucumbers, squash. It can make all the difference for Brussels sprouts or big cabbages, for a broccoli or a thriving pepper bush. The results can be remarkable.

It is important when fertigating that the water sinks right in, making a surface wet spot no larger than the little plate under a teacup. This way the moisture penetrates straight down into the subsoil. The speed the water comes out of the bucket determines how widely the moisture spreads. Exactly how large the drainhole should be depends on the soil type; sandy soil usually accepts moisture rapidly, and the water naturally goes deep and not wide. Clay soils are slower to take in moisture, and it spreads out much more broadly when applied from a single drip source. If the soil has 20 percent clay or more, then five gallons (20 liters) will be about right, given every three weeks. But if the soil has little clay, it has far less ability to hold moisture. In that case, perhaps half as much water, say 2½ gallons (10 liters), given every ten days would work better.

Figure 6.2: *The fertigation bucket at work.*

Fertigation can be a wise practice even when the plants are otherwise getting enough water. If the soil needs no more moisture, you can make the fertilizer solution a bit stronger and give less volume because it will thin itself out with the moisture already present in the earth.

The traditional fertigation solution is manure or compost tea. Fill a barrel (these days it will probably be a garbage can) with water, dump in a bucketful or two of fresh manure or compost, and allow it to brew for a week. Stir the brew every few days. When it's ready, dip out buckets of fertigation concentrate. Periodically refill the barrel with fresh water and add more manure or compost. You'll determine how much you need to dilute this tea according to your results in using it. When the barrel starts getting too full of solids, empty it out, spread the solids on the garden or toss them into the compost heap, and start a new batch. A variation: Put armloads of comfrey leaves in the brew instead of manure. Other forms of organic matter containing a fair amount of proteins that you could toss into the brew include alfalfa (lucerne), seedmeal, or a quart or two (liter or two) of tankage, meatmeal, or even highly potent bloodmeal (this would really put up the nitrogen content). Urine would also work excellently.

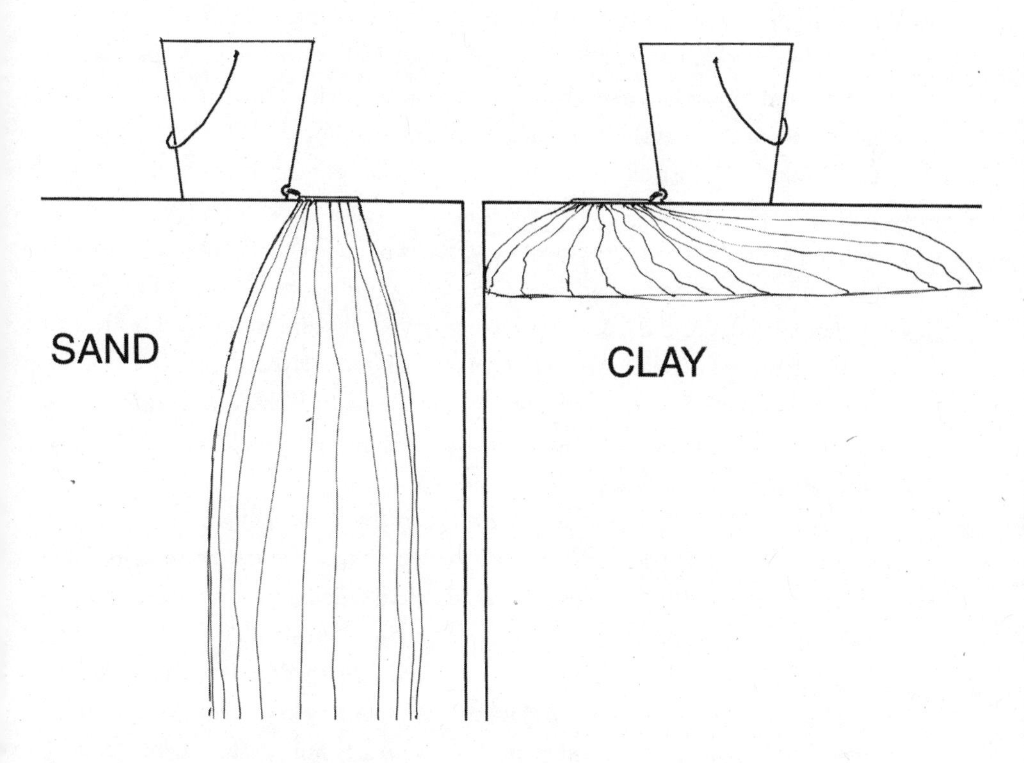

Figure: 6.3: *How water drips into different types of soil.*

These days, someone makes a product for doing everything, and fertigation is no exception. Organic gardeners will see an excellent (and rather costly) result using fish emulsion fertilizer, diluted as suggested, which is usually about one part concentrate to 100 parts water. Soluble chemical fertilizers are also highly effective if they contain trace elements as well as NPK. As long as these chemicals do not replace regular additions of organic matter, their use will not damage soil life because they are applied in a highly dilute form.

Measuring clay content

There is a simple procedure that shows precisely how much clay your soil contains. Every gardener should do this test if for no other reason than to know what their soil profile is to at least three feet (90 centimeters) deep (if it goes that far).

Dig a hole. If you have a soil augur or posthole digger, the hole doesn't need to be any larger in diameter than the tool. Otherwise, neatly excavate a short, hip-wide ditch that is at least 36 inches (90 centimeters) deep. It is normal to find more than one sort of soil below any field. The first 6 to 12 inches (15 to 30 centimeters), the "topsoil" layer, are usually darker in color because they contain almost all the organic matter. Sometimes the same sort of soil particles continue down further, but become lighter in color. Typically, in climates where there is more than about 30 inches (70 centimeters) of rainfall a year, and also in soils that developed slowly out of the rocks they sit atop (upland soils), you'll encounter a change as you dig, almost as though the land were a layer cake. Below some point you will find a layer of clay. In dryer climates this may not occur. If there is such a change, then take a pint-sized (500-milliliter) sample of both types, or all three types if there are three distinct soils as you go down.

Ask yourself a question as you dig: Is this soil loose enough, friable enough, that roots can penetrate it? As you dig are you finding roots? Some soils, especially some clay subsoils, are so dense and airless that almost no roots can exist in them. If you run into that sort of clay, there is little point in going deeper; in fact, in that kind of soil it will be difficult to dig further. If you're gardening on alluvium (soil deposited by flooding rivers or creeks) instead of an upland soil, and if as you dig you run into a layer of nearly pure gravel, there is also no point in going deeper. Few plants will be able to root through it, and gravel serves as a complete barrier to moisture rising to the surface by capillarity — your garden will require considerable irrigation.

So now you've assessed the resource that lies below your feet. I wouldn't want to garden anywhere without having this valuable information.

With samples in hand, the next step is to do a "soil fractional analysis" test. Don't be daunted by the big name. It's a piece of cake. The idea is to determine the makeup of your soil by breaking it into constituent particles and finding out how long they stay suspended in water. The smaller the soil particle, the longer it will stay suspended in water. Clay particles are so tiny that only the most powerful microscope can see one, and they stay suspended in water for many hours, sometimes days. The main thing we want to know is how much clay there is mixed into the larger particles of your soil.

To do the test, get a quart (liter) canning jar or something similar with a good lid. Take about one pint (500 milliliters) of soil, remove any roots or small rocks, break it up as finely as possible, and put it in the jar. Get a felt pen that will write on glass and mark on the side of the jar where the top of the soil comes to. Alternatively, stick a strip of adhesive tape on the glass and mark the top of the soil on the tape. Now fill the jar with water to within about one inch (2.5 centimeters) of the top and add a teaspoonful (five milliliters) of ordinary dishwashing detergent — a low-suds type is best, but low-sudsing is not critical. Seal the jar and shake it hard. And shake. And shake some more. Now shake really hard. Shake for five to ten minutes. Shake until every particle of soil in the jar that had aggregated (formed clumps that resist coming apart) has been separated from every other particle (this is the reason for the soap). If you're lazy or like efficiency, take the jar to the local hardware store and ask them to put it into their paint shaker for a few minutes, then take it home and briefly shake it again. But the second shaking will only take a minute or so because all the soil aggregates will have already been broken up.

Now comes the measurement. Stop shaking and immediately put the jar down in bright light, say on a windowsill. Wait exactly two minutes, by the clock, from the moment you stopped shaking. Then look closely, using a bright flashlight if necessary to see into the murky water. Unless what you've been shaking is pure clay, some of the soil will probably have settled out and will be resting on the bottom. That deposit is your "sand fraction." Mark the depth of the sand fraction on the side of the jar. Now go away for a while, but come back exactly two hours after you stopped shaking, take another look, and make another mark. What settled after the first two minutes but within two hours is the silt fraction. Silt particles are finer than sand but are still just tiny broken and rounded-off bits of rock.

At this point, all that remains suspended in the water is the clay and maybe the organic matter. Clay can take a long time to settle out — days, sometimes weeks. If you wanted perfect accuracy you'd wait until the water was clear and then make another line, but, practically speaking, the first mark you made, showing where the dry soil came to when you first filled the jar, will be close enough. If the clay fraction settles within 12 hours, that's a good sign you have a relatively coarse sort of clay. Any soil that contains much fine clay will be harder to work.

So now you know that you have so many inches of sand, so many of silt and so many of clay. Dredge up your grade-school math and calculate the percentage of each. (This calculation is far easier when measurement is done in millimetres.) For those readers whose math is weak, you can determine the percentage by dividing the height of the dry soil into the height of the layer in question.

Run this test on each soil layer you encountered when digging your hole.

One benefit of knowing the percentage of each type of soil in your garden is that the mix of soil particles indicates how much water that soil can hold for the benefit of growing plants (see Figure 6.4).

Water retention by soil type

Soil type	Total moisture-holding capacity (inches per foot)	Amount of water lost when most veggies begin to experience moisture stress (70% of capacity) (inches per foot)	Amount of water still held when plants wilt and die (inches per foot)
Sandy	1.25	0.5 to 0.75	0.25
Medium	2.25	1	0.60
Clayey	3.75	1.5	1.30

Figure 6.4

A sand soil is defined scientifically as one with less than 10 percent silt or clay. A medium soil could be any mix of sand and silt that has at least 20 percent clay but less than 35 percent clay; this sort of soil is commonly called a "loam." A clayey soil is made of more than 35 percent clay. If it contains no more than half clay, it can be an excellent soil to grow a garden in. If it is more than half clay, it can be a difficult soil to work.

Okay, now you've got some solid information about your soil. Especially important, you've got an idea of how much moisture it will make available to plants. A sandy topsoil with a clayey subsoil that is open to root penetration can make more water available than it might seem when you just consider the top few inches. If the subsoil is a deep, open clay, there will be a huge reserve of moisture. On a soil like this you may see plants wilt a bit in midafternoon during the hottest weather, but they will seem restored the next morning (although they aren't entirely restored because any wilting is a shock to a tender vegetable plant).

Wise gardeners do everything possible to help their plants avoid moisture stress and especially to avoid wilting. If you're not irrigating at all, then by increasing spacing you give each root system a greater area of soil to mine for

Amount of water lost per day in midsummer

Temperature in summer	Region	Inches lost per Day
Cool	Western Washington, coastal southern B.C., Canadian Maritimes, Scotland, and western U.K.	0.2
Moderate	Western Oregon, northern US, Southern Canada, Tasmania, New Zealand (S. Is.), eastern U.K.	0.25
Hot (humid)	Eastern US, east coast Australia, NZ (N. Is.)	0.3
Hot (dry)	Prairie states, northern California, Coastal southern California, southwest Oregon	0.35
Low desert	Southwest desert states, southern California	0.45

Figure 6.5

water and give capillary uplift a greater chance to effectively recharge your plants' root zones. If you are fertigating, your soil fractional analysis will tell you if you should add water in small increments more often to sandy soil or if you can add water in larger amounts less frequently because there is enough clay to hold the moisture. If you are irrigating regularly and spacing your plants according to Column 2, you will need a bit more information to irrigate most effectively.

Figure 6.4 showed how much water loss you should allow before irrigating to prevent moisture stress. The amount of water that is extracted from soil each day depends on the air temperature, the humidity of the air, the strength of the wind, how dense the crop canopy is, etc. But as a rough guide, Figure 6.5 shows, roughly, the amount a crop that has made a canopy will draw up and evaporate during midsummer.

Suppose I have the gardener's dream, an infinitely deep loam. I have plenty of irrigation water, so I have used Column 2 spacings. After I bring it to its water-holding capacity, my soil can give up about an inch (2.5 centimeters) of water before the crops will experience any moisture stress. If I can prevent moisture stress, the garden will grow better and produce more. Suppose I am gardening in Maryland and it is a hot, sunny, rainless week in mid-July. Most of my vegetables have formed a dense leaf canopy, so my ground is losing about 0.3 inches (seven millimeters) per day. I must add about an inch of water to the soil every three or four days. Of course, if there is a spell of cloudy weather, much less water will be lost. If it drizzles insignificantly, at least during that time, no moisture will be lost from the ground. If it rains, I can add that amount to the moisture balance of my soil. But if I owe the garden an inch of water, and just before I irrigate it rains two inches (five centimeters), this account isn't the type that can go into surplus. Any extra moisture will move into the subsoil and can only assist the plants after rising slowly by capillarity. So even though I just got two inches of rain, when the soil has lost an inch a few days from now, it will still be time to add another inch.

Foliar feeding

When I was doing dry-gardening experiments, I noticed that if, instead of fertigating, I sprayed liquid fertilizer at the same strength with which I'd fertigate on the leaves of plants that looked as through they were moisture stressed, then the plant almost instantly looked quite a bit better. Not as much better as it would had I fertigated, but still, considerably better. This practice is called foliar feeding. If you are so short of water that finding a few hundred gallons a week for fertigation is not possible, then perhaps a few gallons a week mixed into the sprayer would be the next best choice.

A few hints: If you're going to foliar spray homemade teas, they need to be perfectly filtered first. Farmers in India have discovered that one of the best possible foliar sprays is half-strength Coca Cola. It contains significant amounts of phosphoric acid, and there is something about the sugar that helps plants a lot. If you should encounter a half bottle of flat Coke, why not give it a try?

Regular foliar feeding is not a bad practice in any case. Kelp tea is one of the best organics to use for this purpose when the plants don't need more nitrates. Kelp contains phytamins and a full range of trace mineral nutrients. A mixture of kelp tea and fish emulsion sprayed once a week will make the plants grow much faster. So will a complete soluble chemical mix if it has trace minerals in it too.

A gardener's textbook of sprinkler irrigation

If you've designed an intensive garden that depends on regular irrigation, or if you plan to irrigate as an emergency measure in the event of a longer-than-usual dry spell, this section will help you become systematic about watering. You need a system based on science because both watering too little and watering too much can hugely reduce the garden's performance.

It soon becomes obvious to the new gardener that even short periods of moisture stress reduce vegetable quality. This stress does not have to reach the point of wilting to cause a lot of damage. Moisture stress can happen to regularly irrigated gardens, too. Underwatering may be so subtle that no temporary wilting occurs, thus going unnoticed, but it still produces serious consequences.

Many gardeners wet down their gardens almost daily with hose and fan nozzle because they want to make sure the beds provide abundant moisture all the time. This method, done properly, works quite well, so long as you enjoy the task. However, if the garden is watered daily but too briefly, the plants can become water-stressed, although on casual inspection the soil seems moist. What is happening is that the surface inches stay damp, but the gardener never discovers that, deeper down, the soil has become bone-dry. Under these conditions vegetables will not wilt, because they never quite suck the surface layer dry, but they'll become severely stunted due to lack of root development.

Recognizing this possibility, John Jeavons, the "father" of intensive gardening in North America, in his book *How To Grow More Vegetables ...*, recommends using a fan nozzle daily, continuing on each bed until the entire surface sparkles (becomes shiny wet). This sparkling results from water that has not yet flowed into the bed beading up on the surface. The shine only lasts a second or so initially, but as the deeper soil becomes saturated, the shine lasts longer and longer. Jeavons says that when it lasts long enough (one to ten seconds), the bed has been given enough water.

One to ten seconds is quite a range! Jeavons suggests trying different shiny times, combined with digging some test holes to see how deeply the soil is saturated. Without this check, gross overwatering or underwatering could result. Many clayey soils, capable of holding a great deal of water but often assimilating it slowly, could accept a lot of water before it penetrated far. So clays could show a sparkly surface for several seconds, yet be quite dry a few inches down. Coarse-textured sandy soils usually take in water rapidly, and it might be difficult to get a shine to last more than one second, if at all, no matter how much the gardener overwatered or how fast the water was put down, resulting in considerable leaching.

Then there are types of fine sands that become coated with the products of humus breakdown in such a way that, once they have become dry, water beads up on them and runs off without penetrating. It is very difficult to remoisten soils like this after their surface has dried out. Applying high-volume flows with a hand nozzle doesn't work very well on soils like this. Gardeners must either mulch these fine sands so the surface never dries out, or make a small investment in low-application-rate crop-sprinklers (more about these shortly).

When I was a novice gardener, I hand-watered as Jeavons suggests. However, with businesses to run, I needed a less time-demanding method. I found that overhead sprinkler systems could achieve about the same result without consuming my time. If I had to be away, I could ask my wife to please turn on all or part of the system and run it for a specified time. But it is perfectly possible to water quite successfully with no more equipment than a fan nozzle and a hose — so long as you know what you are trying to accomplish and fully appreciate how much water your soil will take in when sprayed with a nozzle for a given amount of time.

What a sprinkler system should accomplish

When you irrigate or when it rains, each soil particle attracts to itself all the water that will stick to it before the force of gravity overcomes this attraction and pulls surplus water deeper into the ground. Thus, the surface few inches of soil can quickly become saturated, while deeper layers may still be dry. A layer of soil that has absorbed all the water it can hold against the force of gravity is said to be at *field capacity*. It's like a sponge retaining all the water it can. After saturation has been reached, if more moisture is added, some starts dripping out the bottom. Continuing the irrigation brings layer after layer to capacity, and the moisture seeps ever deeper. Each and every irrigation also leaches water-soluble plant nutrients out of the surface layers and moves them down to the full depth the water has reached. If you water too long, plant nutrients are moved so far down that your vegetables' roots can no longer access them. Result: poor growth.

The opposite of soil at field capacity is totally dry soil, something rarely — perhaps never — seen on this planet. That's because as soil particles dry, the moisture they're holding becomes an ever-thinner film on their surfaces. The thinner the film, the more tightly it is held, until moisture clings so tenaciously to soil particles that vegetable roots can't extract it. If soil gets dry enough, the remaining moisture clings so hard that evaporation at normal temperatures can't remove it. To get soil completely dry, you would have to heat it to exceed the boiling point of water. Even the hottest, never-rained-here-ever desert soil still holds a little moisture.

The point on a wet-to-dry scale where moisture clings so hard to soil particles that plants can no longer extract any moisture is called the "permanent wilting point." Most vegetable species are not very effective at extracting moisture from dryish soil and wilt permanently at the point soil is holding about a third of the water it potentially could hold. But well before the permanent wilting point is reached, the soil comes to a degree of dryness at which vegetables experience temporary wilting during the few hours the hot midday sun increases their need for water beyond the ability of their roots to extract enough. Always imaginative, soil scientists call this degree of dryness the "temporary wilting point." And well before their soil gets even that dry, most kinds of vegetables, being fragile, highly inbred, weakly rooting creatures, begin to experience subtle, almost unnoticeable moisture stress. It is your job to prevent this stress; as I've said before, any stress reduces the quality and amount of production.

How much to water

Plants convert solar energy into new plant material. If photosynthetic efficiency could be increased, plant breeders could create new varieties capable of producing far more plant material in far less time. So far, this has proved impossible. As a result, when plant breeders seek to increase crop-plant productivity, about all they can do is redirect the focus of a plant's efforts — emphasize one aspect of a plant's activities by encouraging it to put less energy into some other area.

Vegetables are far less able to survive moisture stresses than field crops like wheat are, because vegetables have been trained over millennia to make larger edible parts at the expense of root system vigor. In the last 50 years or so, vegetables have been bred to become even weaker in this respect because to maximize profit in our industrial age of oil-driven irrigation, modern plant breeders have redesigned many vegetable varieties so they will produce even larger edible portions even more quickly at even greater cost to their root development.

Modern industrial farming aims at maximum production by maintaining vegetable field moisture levels above 70 percent of capacity, to the full depth of the vegetables' root development. Intensive raised-bed gardeners should try to accomplish about the same thing. Once the top 12 inches (30 centimeters) of soil has dried to about 70 percent of its moisture-holding capacity, it should be watered back up to capacity again. By the time the top foot has dried to 60 percent of capacity and the next foot has dried to close to 70 percent, the bed *must* be watered back up to capacity.

If you'll contemplate Figures 6.4 and 6.5 and do a little arithmetic, you'll see that in moderate climates it would be wise to water semi-intensive raised beds (Column 2) back up to capacity every two to five days — more frequently (giving less water) on sandy soil; less often, but applying more when you do water, on heavier soils. The basic plan should be to replace lost moisture more or less at the rate it is lost, without overwatering (which leaches out soil fertility).

Judging by recommendations in garden books and magazines, and judging by the equipment most gardeners use for irrigation, I conclude that gardeners who do irrigate grossly overwater far more often than they underwater. Consumer-grade "lawn and garden" sprinklers spread water thick and fast. Although consumers don't get accurate performance specifications, as buyers of agricultural-grade crop sprinklers do, it is easy to test any sprinkler for application rate. Simply set out a few water gauges — empty tin cans or other cylinders with straight-up-and-down sides. Put one near the sprinkler, one near the outer limit of its reach, and a couple in between. Run the sprinkler for exactly 20 or for exactly 30 minutes, measure the depth of water in each container, average those amounts, and do some arithmetic to derive the sprinkler's "application rate per hour." Most lawn sprinklers spread well in excess of two inches (five centimeters) per hour. Oscillating sprinklers, the kind that spread water in rectangular patterns, put down two to four inches (five to ten centimeters) per hour. The exact amount depends on sprinkler design, water pressure, and pattern adjustment knob setting. Spraying or soaker hoses and spot sprinklers designed to cover small areas usually put out an even higher rate. How much leaching do you suppose the average gardener causes by running one of these sprinklers for only one hour? (Please contemplate Figures 6.4 and 6.5 again.)

Another benefit of doing this water-gauge test is that you'll see the uniformity of distribution (or lack of it) that you're getting. Try it! You may be saddened by your results. However, any sprinkler that wets the ground fairly uniformly, even a high-output one, can water a garden effectively without leaching if you know the sprinkler's application rate and know how thick a layer of water you wish to spread.

Now you need to determine *when* to water. Soil moisture is best judged five to six inches (12 to 15 centimeters) below the surface. Firmly squeeze a handful of soil into a ball — the classic ready-to-till test. If the ball feels quite damp and sticks together solidly (unless you have very sandy soil, in which case this test won't work at all), the soil moisture is above 70 percent. If the moist soil ball sticks together firmly but breaks apart easily, the moisture content is around 70 percent. If the soil feels damp but won't form a ball when you squeeze it hard in your fist, and if the soil contains over 10 percent clay, your vegetables are experiencing moisture stress. They may look okay, but they are still at least mildly stressed.

Another, more convenient, way to determine when to water is to knowledgeably guesstimate the amount of moisture recently lost from soil. The amount of water that sun, wind, and heat remove from soil varies with the season and the amount of vegetation the soil supports, but does not vary with the type of soil. Regardless of their texture, all soils lose water at about the same rate because it is not the sun shining on the earth that dries soil out; it is the sun evaporating moisture from plants' leaves.

If you're going to base your additions of water on the figures in Figure 6.4, remember to adjust for cloudy days (when far less loss will occur) and for any rain received (even if it is only a drizzly fraction of an inch that stops the soil from losing moisture). Reduce the rate of loss for areas that aren't covered by a leaf canopy — bare soil hardly loses moisture at all, especially if it has been cultivated and has formed a dust mulch.

Sandy gardens should be irrigated after losing about half an inch (1.25 centimeters) of water so as to reduce leaching. Clayey soils growing larger vegetables with two-foot-deep (60-centimeter) root systems can easily accept 1 to 1½ inches (2.5 to 4 centimeters) of water without danger of leaching. You may need to apply more frequent, lighter irrigations on heavier soils to keep the surface moist when seed is sprouting, when you are nursing recently transplanted seedlings during hot weather, or for species with unusually high moisture requirements, such as radishes and celery. These extra needs are best supplied with a hose and fan nozzle.

Designing sprinkler systems

Agricultural-grade sprinklers spread water more uniformly than the consumer-grade stuff home gardeners usually use. Agricultural sprinkler heads spread water at a known, specified rate and are *not* designed to wear out quickly, as too many consumer-grade ones are. In the long run, paying whatever is necessary and travelling as far as necessary to a supplier of commercial-grade equipment works out to be far less expensive. For the nearest supplier, check the Yellow Pages under "Irrigation."

Your choice of sprinkler size can make quite a difference. Although it takes a larger number of low-application-rate sprinkler heads to cover a given area (small-bore nozzles, shorter throw radius), it is better to use this sort in the home garden because (1) they put out lighter droplets, and (2) a smaller throw radius helps keep the water where you need it and off adjoining buildings and noncritical vegetation. High-application-rate sprinklers put out large heavy droplets that can cause significant soil compaction, reducing root development and making cultivation and weeding more difficult. Large droplets pounding on the surface may also contribute to forming a soil crust. Because crusts don't form on sod, and because the lawn itself breaks the force of large

Comparison of high- versus low-application-rate sprinkler performance

Nozzle diameter in inches *	Operating pressure in PSI **	Discharge in gallons† per minute	Throw radius in feet ††	Spacing in feet	Application rate in inches per hour
1/16	30	0.45	33	20 x 20	0.11
1/16	60	0.79	36	20 x 20	0.19
7/64	30	1.94	33	20 x 20	0.47
7/64	60	2.66	36	20 x 20	0.64
13/64	30	6.78	40	25 x 25	1.05
13/64	60	9.53	45	25 x 25	1.46

* Multiply by 2.54 for centimeters.

** Multiply by 7 to obtain kpa (kilopascals); multiply by 1.42 to obtain meters of head.

† Multiply by 3.785 for liters

†† Multiply by 30.48 for centimeters

Figure 6.6

droplets before they hit the ground, most lawn sprinklers issue big droplets and produce high precipitation rates — apparently a timesaving convenience for busy homeowners, even though sod leaches as easily as vegetable plots, with the same consequences. Consider how much overwatering can happen if one of these monster sprinklers runs forgotten for a few hours.

Sprinkler systems that apply less than half an inch (1.25 centimeters) per hour are best for the garden but do have drawbacks. Sun, wind, and high air temperatures can combine to break up fine streams of water and evaporate them nearly as fast as the sprinkler puts moisture out. Most criticism of sprinkler irrigation coming from drip enthusiasts (or drip equipment salespersons) is based on this misuse of the method. Small sprinkler heads are not useless, but you should not use them when the sun is strong and the wind is blowing. Sprinkling when it is calm and the sun is lower or hidden by clouds is more efficient, too, because little or no water is lost before it enters the earth.

It is possible to design a system with such a low application rate that you can water a clay soil all night long, from bedtime to breakfast, without risk of overwatering. At night there is rarely any wind, and for rural homesteaders with limited-output wells, this is also the time when there is no competition from showers, dishwashing, and so on. Watering at night is widely believed to be harmful to plants, but it may actually be the best time to water if your heavy soil allows you to do it all night without leaching. Plants are naturally dampened by dew for the last few hours of the night; during summer they quickly dry off in the morning. You *can* harm plants by watering them just before dark, stopping the irrigation, and leaving the plants damp all night. These are ideal conditions for the multiplication of disease organisms. Watering all night continuously washes bacteria and fungus spores off the plants before they can do any damage. This principle is well understood by nurseries that propagate healthy plants from root cuttings under a continuous mist.

On lighter soils, the best sprinkler design is one that spreads about a quarter of an inch (six millimeters) per hour, allowing you to apply a layer half an inch to three quarters of an inch (12 to 19 millimeters) thick when you sprinkle for a few hours in the early morning before breezes start up and the sun gets strong.

Although nozzles even smaller than the ones noted in Figure 6.6 are available, 1/16-inch (1.8-millimeter) bores are about the minimum effective size for veggie gardening. Systems designed around this nozzle size cover the largest amount of ground while using the smallest possible number of gallons per minute to do it. Bores with a smaller diameter can't spray far enough to achieve even lower application rates, so you need to use many more sprinklers to cover the same area. This gets expensive.

As the bores in nozzles get larger than 7/64 inch (2.75 millimeters), they start emitting droplet sizes too massive for a veggie garden. These might be fine for pastures, golf courses, or corn fields, but to run several of them at once requires a larger water supply than most home gardeners have available. Large-scale farmers routinely use sprinklers with nozzles of 13/64 inch (five millimeters) and larger, each sprinkler drawing in excess of five gallons (20 liters) per minute.

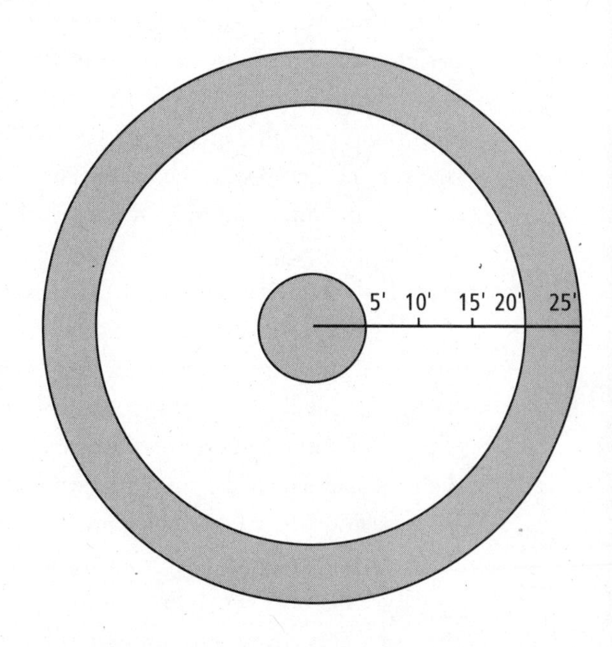

Figure 6.7: *The formula for the area of a circle is: $A=\pi r^2$. Imagine a sprinkler with a 25-unit radius (for the convenience of American readers, assume that the "unit" is feet, but it can be any measure, metric or imperial). The innermost five units of the sprinkler pattern occupies 78.5 square units (3.14 x 5 x 5). The area in the outer five feet of the pattern is 706.5 square units, calculated this way: the area of the full circle (3.14 x 25 x 25) minus the area in the inner 20 feet of the pattern (3.14 x 20 x 20). Thus the nozzle must deposit nearly ten times as much water on the outermost five units of the pattern to end up with the same thickness of coverage as on the innermost five units.*

Sprinkler designs. It seems to be impossible to design the perfect crop sprinkler, a single sprinkler that uniformly spreads water over a circle or a square while it sits in the middle of the space. The water-gauge test described a couple of pages earlier will show you how difficult it is to solve this problem. Use any sprinkler made by any manufacturer you care to choose — whether lawn, garden, agricultural, or commercial. Not only does no single sprinkler I know of accomplish perfectly uniform coverage, but most fail miserably. The reason? With a circular pattern, a sprinkler must deposit nearly ten times as much water on the perimeter as it does in the center. Every point in between receives a different amount.

Many design tricks are used with agricultural sprinklers to approach the ideal of equal water distribution, but even the best of them probably puts twice the amount of water on the inner half of its coverage. Surprisingly, the design that seems to so cleverly overcome this problem by watering in squares and rectangles instead of in circles — the oscillating sprinkler — is usually the worst culprit of all. I suspect the reason for this is that the cam arrangement that rotates the spray arm is inevitably a loose fit that pauses the arm too long at the turnaround point, so this type of sprinkler errs by putting too much water at the ends of its rectangular pattern and too little above the sprinkler itself. Do you have one? Measure it yourself!

The impact sprinkler can't apply water uniformly because the rocker arm passing through the nozzle jet (its bouncing rotates the sprinkler head) dumps too much water close to the sprinkler while too little is thrown to the extremes. Most consumer-quality impact sprinklers come with a diffuser paddle or adjustable needle-tipped screw of some sort to shorten the water throw by diffusing the spray. But more than the slightest amount of diffusion increases the tendency to grossly overwater the center while leaving the fringes too dry. The more the radius is shortened by breaking up the nozzle stream, the worse this effect becomes. Agricultural-quality impulse sprinklers do not use diffusers; instead, they have nozzles with scientifically designed bores that, if used at the correct pressure, diffuse the stream (spray) properly all by themselves, putting only about twice as much water near the center of their coverage as on the outer half. Again, if you have one, measure it yourself!

To compensate for this inherent limitation of sprinkler design, farmers set out many overlapping sprinklers, all going at one time. These are arranged in regular geometric patterns so that one sprinkler's heavily watered area is overlapped by another sprinkler's deficiently watered area, and the differences roughly average each other out. Any multiple-sprinkler pattern still leaves a dryish fringe area, where fewer overlaps occur. On the farm, these fringes beyond the margins of the field are of little consequence; in the backyard, it may be essential to keep fringe areas within your own yard, if only to keep overspray out of neighbors' yards or off windows.

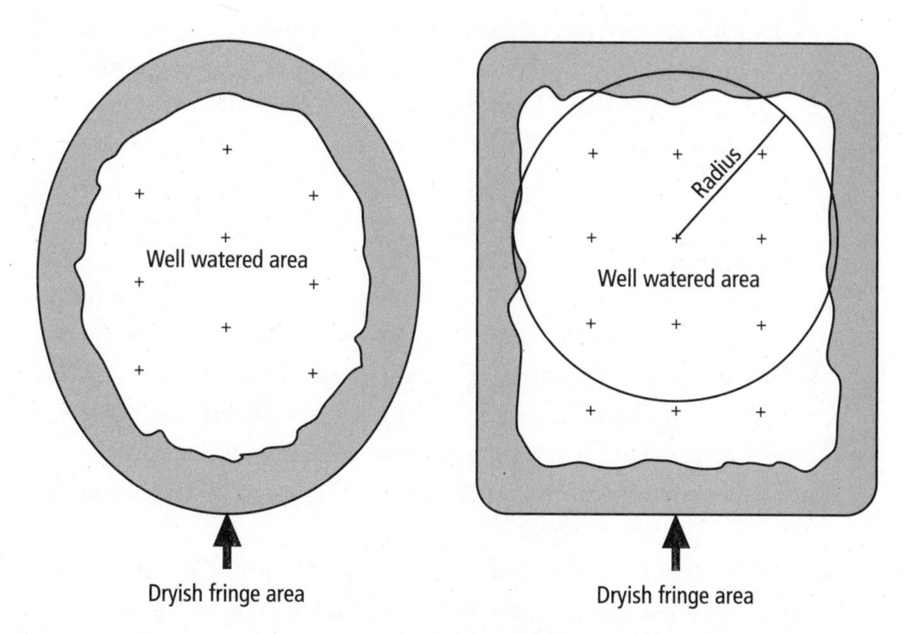

Figure 6.8: *Simultaneously operating several correctly spaced sprinklers creates overlapping water patterns, permitting uniform coverage. The optimal spacing between sprinklers is about 60 to 70 percent of the sprinkler's throw radius. Left side: Sprinklers arranged in triangular patterns. Right side: Sprinklers correctly overlapped, arranged in square patterns.*

The less-watered fringes can, however, be useful for growing a dry garden or for locating tall perennials, especially if you're using low-angle sprinklers (to be explained shortly). Putting a line of raspberries or climbing beans right along the beginning of the dry fringe intercepts all the overspray so that it drips off the leaves and concentrates in this row. In this way, the hedge gets about as much water as is deposited in the middle of the sprinklers' patterns, and nothing much goes past it.

Sometimes sprinkler patterns are laid out in squares, sometimes in triangular patterns. The triangular method spreads water slightly more uniformly, but the square pattern may lend itself better to the backyard situation. The shorter the designed radius of the sprinkler is, the smaller the fringe area will be, making short-radius, low-angle sprinklers preferable for backyard gardens.

Probably the single most useful, inexpensive, and highly durable agricultural-quality impulse sprinkler available is the Naan 501, made in Israel. It comes with assorted nozzle sizes, designated in metric. The best are the 1.8 or 2.0 millimetre nozzles, which is close to 1/16 inch. Naans are available through many big irrigation and farm suppliers, who can order them through their wholesale sources if they don't actually stock them. I have used Naan sprinklers since the early 1980s. The best of the Naan range is designated as model 501U, because the 501 all by itself is nothing but the sprinkler head, which requires supports and feeder systems for proper installation. The cost of all these bits exceeds the price of the sprinkler. However, the "U" model comes with a one-meter-tall (slightly over one-yard-tall) metal support rod that holds the sprinkler head well above most vegetable crops, and a feeder pipe with a seven-millimetre (about ¼ inch), quick-disconnect barbed fitting at the end — all ready to go. If the supplier you contact does not have the 501U in stock, don't let him or her fob off the lesser items on you. Demand a special order and wait a few weeks until it comes in.

If you are using these, I strongly suggest you also purchase a special seven-millimetre punch to insert the barbs into the supply lines. If you make these holes with a nail, they'll leak. The punch only costs a few dollars and will last a lifetime.

Some gardeners try to eliminate fringe areas by using impact sprinklers with part-circle mechanisms, locating sprinklers at the edges or corners of the garden. Generally this practice has more disadvantages than the gardener realizes. Keep in mind that cutting the arc in half doubles the rate of application; cutting it to 90° quadruples the rate of application. Part-circle sprinklers have two other flaws. First, you need a rather large-bore nozzle to create enough force to actuate the part-circle mechanism, and such a nozzle puts out a powerful stream with high application rates and big droplets. And second, when the sprinkler's mechanism is in its reverse cycle, it dumps a lot more water close in, further accentuating the undesirable tendency to overwater close in.

The best (but not cheapest) alternative to the impact sprinkler was invented by the Toro Company. This design is now made by several other companies too, including an Australian manufacturer. It uses an internal water-powered turbine to rotate the sprinkler head. This type also has multiple-nozzle heads with various emission rates that provide the most uniform coverage possible and, interestingly, the ability to cover part circles without increasing the application rate. This design was initially intended for institutional situations, where sprinklers have to be located close to buildings and windows and where the user wants precise coverage in order to avoid sidewalks, overspraying windows, and so forth. Gardeners with fat wallets and an interest in high-tech irrigation should consider these.

High- and low-angle sprinklers. Agricultural-grade sprinkler heads are designed with different angles of throw. High-angle nozzles allow the stream of water to go its maximum distance, covering the largest area with the fewest number of sprinkler heads while the entire system draws the least number of gallons per minute, resulting in lower application rates. However, high-angle sprinklers are more strongly affected by wind, which can disperse the water stream, blow it off course, and cause high evaporation losses, especially if the sun is shining. High-angle sprinklers can be a wonderful solution for homesteaders who want large gardens but have low-yielding wells — if they avoid the sun and wind by watering at night or very early in the morning.

Low-angle sprinklers are best for windier positions or for daytime use. Low-angle sprinklers throw at only a few degrees above horizontal, so the radius is shortened and the stream is kept close to the ground, out of the strongest wind gusts. They're better in tight backyard situations, too. More low-angle sprinklers are needed to cover a given area, resulting in somewhat higher precipitation rates. The Naan 501 series sprinklers have low-angle nozzles.

The fine points. When gardeners first study a commercial irrigation catalog, they sometimes become confused. Here are a few hints if you want to become educated by it instead.

Agricultural sprinklers come with recommendations for spacing and operating pressure. Operating a sprinkler outside its design limits results in poor performance. In the case of crop-sprinkler nozzles, matching the shape of the nozzle's bore to the water pressure is especially important. If the water stream is propelled from the nozzle by a pressure too low for the design, the stream (jet) doesn't break up and spray (or diffuse) properly. The impulse arm, as expected, causes much water to be laid down near the sprinkler, and the tight, undiffused stream carries water to the fringes, but few droplets are landing between these extremes — little water is laid down in the middle of the pattern.

Consider the opposite effect. Run at excessively high pressure, the water jet mists and breaks up too much — "sprays too much," as a farmer would say

— actually shortening the throw of the water, greatly increasing the rate of application near the sprinkler, and making the more distant parts of its coverage too dry.

Nozzles are designed to spray properly at pressures ranging from ten to 100 pounds per square inch (700 kpa), with most requiring from 30 to 60 psi (210 to 420 kpa). Once piped water gets past your house's pressure regulator, its pressure is usually between 30 and 45 psi (210 to 315 kpa). The pressure in unregulated water mains can be considerably higher or, unfortunately, sometimes lower than this. Lucky homesteaders having their own electric pumps can, within limits, choose their water pressure. *Do not attempt to use sprinklers demanding higher pressure than you have.*

High-angle sprinklers should not be spaced at more than about 65 percent of their radius. This allows the proper amount of overlap in the pattern and allows for wind blowing the spray a little without making areas dry. Low-angle sprinklers are less bothered by wind and are usually spaced at 75 percent of their radius.

Buy agricultural sprinklers from a farm supply or irrigation company. If you live in or near a city, you can also buy them at landscaping firms, although you'll probably have to chose the ones you want from a catalog and have them ordered in. These suppliers usually have a wide range of flexible plastic irrigation pipe and quick-connect fittings as well, so you can design a supply system that will handle the amount of water flow required. The dealer should be able to offer lots of good advice on how to assemble such a system.

The trickiest part of designing a sprinkler system is improvising the risers that hold the heads. In the last two decades, sprinkler-head stands have been made of inexpensive black plastic. The trouble with most of this plastic stuff is that it is designed for lawns and ornamental borders, so the risers are not high enough to hold sprinkler heads above vegetable crops. Veggie gardeners need supports standing about three feet (90 centimeters) above the soil. (As mentioned earlier, the Naan 501U, atop a tall metal spike, takes care of this for you.) One way to improvise risers is to glue a cheap, short, plastic sprinkler spike into the end of a long piece of heavy, three-quarter-inch (two-centimeter), white plastic pipe. Cut off the bottom end of the pipe at a sharp angle so that you can push it into the soil. The sprinkler spike is supplied through a barbed push-in connector in a black ABS plastic supply line laid atop the soil; the white plastic pipe holding up the sprinkler spike carries no water. If all this seems hard to imagine, spend some time at a supplier's, handle the bits sold there, and see if all the pieces don't fall into place.

Gardeners who know what they want to accomplish can reach their goal without ideal equipment. If a complete, permanent, multiheaded sprinkler system that turns on from a single valve is beyond your interest or budget, you can still achieve uniform irrigation with one good sprinkler head on a tall stand supplied by an ordinary hose. Move it around the garden and run it for

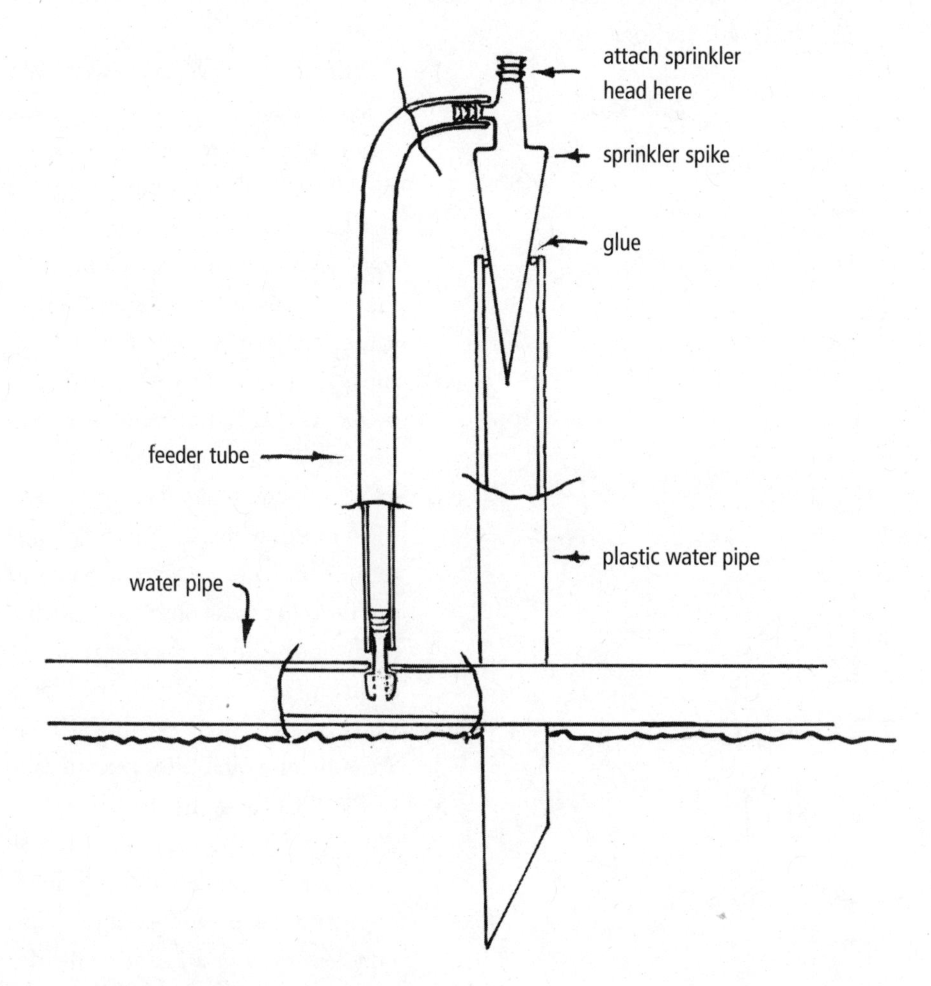

Figure 6.9: *An improvised sprinkler-head riser.*

equal periods in carefully determined positions. I did it this way for the first few years I ran a trials ground (before my seed business made any serious money). I made a stand with a sack of ready-mix concrete, a five-gallon (20-liter) plastic bucket, four feet (120 centimeters) of galvanized pipe, and a few fittings.

Drip systems and microirrigation. I do not recommend drip systems for the home garden. I used drip tubes on my trials grounds from 1982 until 1986 simply because drip was the only way I could water extensive areas during daylight hours (my property had a puny well supplying a bit less than three gallons [11 liters] per minute).

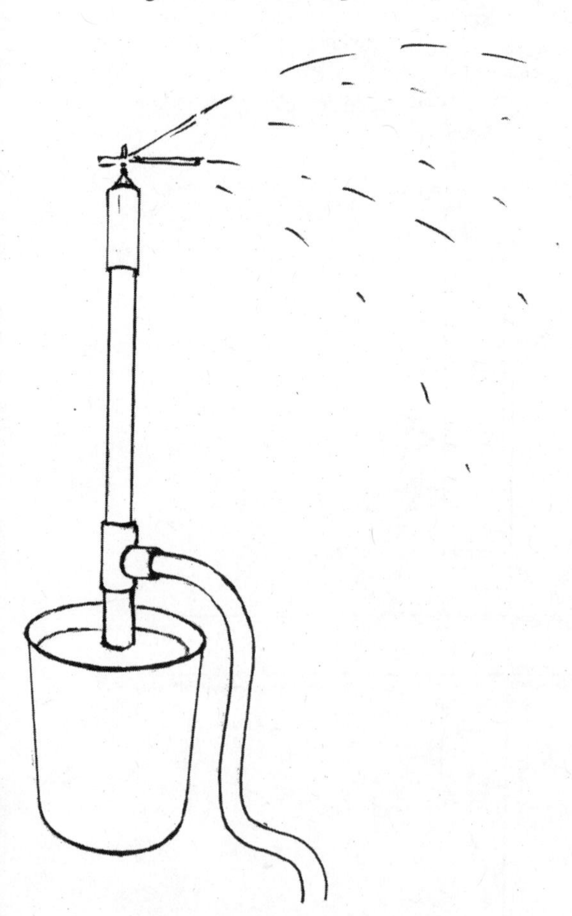

Figure 6.10: *An improvised single-head sprinkler system.*

Drip tubes are expensive, even when purchased in 1,000-yard-long (around 900-meter) rolls. They are also short-lived and troublesome, but at that time I did not care what it cost in money or effort to produce my trials grounds — I was growing valuable information, not food. Drip tubes are easily cut with sharp hoes or shovels, and emitter holes tend to become plugged up at times, even if you have water filters. This means you must carefully inspect the entire system each and every time you turn it on. Drip lines also shift many inches from side to side as they expand and contract, so they won't dependably water a line of new seedlings. They are even less suitable for germinating seeds. Drip systems are absolutely not workable on sandy soils because the water goes straight down through sand without spreading out horizontally, leaving large areas of totally dry soil that the plants can't root into.

High-quality long-lasting drip lines might be useful for permanent plantings, such as raspberries in heavier soils, but given a choice between drip systems and sprinklers, I'd always choose sprinklers.

Lately, new advances in plastics manufacturing have created a hybrid between drip and sprinkling, called microirrigation. These systems use inexpensive low-pressure plastic tubing to carry water; cheap quick-disconnect fittings for corners, plugs, connectors, and tees; cheap plastic spikes to hold sprinklers; and miniature short-radius sprinkler heads with emission rates so low they are measured in gallons (or liters) per hour, not per minute. Microirrigation systems provide an inexpensive and durable alternative under orchard trees and in vineyards. They are also being used more and more by homeowners to water ornamental beds around houses, and they are very useful in tunnel cloches to keep plants watered for a few weeks until the cloche is removed. Microirrigation bits come bubble-packed (like screws and bolts) in garden centres, but if purchased that way they are expensive compared to what you'll find among the much broader assortment housed in the shelf bins of agricultural suppliers. If you are considering a microirrigation system, be wary about getting uniform water application. And use very effective filters! The nozzles are extremely fine.

CHAPTER 7

Compost

In Chapter 2 I suggested using a soil-improvement program based on purchased amendments. But isn't this a book about raising food without spending much money? I have three reasons for saying what I did in Chapter 2: (1) because as I write these words in 2005, these materials are still inexpensive and readily available; (2) because when people first start a garden they likely won't have had time to make compost; and (3) because even with an ongoing composting operation in place, using complete organic fertilizer (COF) or something like it is still an excellent practice.

I repeat: feedlot manures and composted chicken manure are widely available and seem inexpensive to most people *right now*, and industrial agricultural wastes like seedmeals and slaughterhouse wastes are also cheap. But if present trends — peak oil, climate change, irresponsible monetary manipulations by central banks, competition over resources — result in the least desirable outcomes, then ordinary people will find it ever more difficult to afford to eat healthfully. It seems almost inevitable to me that the real (inflation-adjusted) cost of soil amendments is going to increase substantially. Composting is the alternative to purchasing. It allows small-scale food growers to make their own fertilizer, to manufacture "well-rotted manure" without the need to own livestock.

I need to make three brief apologies before getting into the substance of this chapter. I write by speaking intimately to imaginary people sitting in invisible chairs next to my computer monitor. I always envision these people as friendly, accepting, understanding. And for this book most of my imaginary listeners are new, inexperienced gardeners, feeling a bit bewildered. Up to this chapter I have tried to keep it simple, starting them off on the right foot so they'll discover by observation the rest of what they need to know. But the subject of composting is different. Making low-grade compost suitable for mulching ornamentals or fruit trees is not difficult. But producing compost that will effectively grow vegetables is a highly skilled activity requiring hard work, close attention, and a desire to "see" what is happening inside the heap. This chapter cannot be "made simple," and for that I apologize.

And the second apology. The subject of composting is a huge one. Fifteen years ago I wrote a book as big as this one that was only about composting. There is no way I can talk about every aspect of composting in one brief chapter. The techniques I am going to describe in this chapter are the ones most likely to result in success for the new composter. If you want to learn the whole of compost making, there are heaps of books in print on the subject, but be prepared to sort out a lot of contradictory data.

Finally, please accept my apology for not lying to you. Many garden writers present every kind of composting as being inevitably successful and easy, portraying the result, always, as effective fertilizer. This is not actually the case. There are many composting pitfalls. If you can't afford to have your food gardening efforts fall far short of your hopes, if you can't afford to just shrug off this year's catastrophe and hope it'll be better next time, you need the truth.

For all these reasons I suggest that you read this chapter more than once.

I compost only to recycle garden and kitchen waste. Because I never end up with enough of my own compost to satisfy my garden's need for humus, I also use well-decomposed feedlot manure, which is another form of compost. I buy feedlot manure because, like my own compost, it gives the soil better tilth and provides food for a healthy, stable ecology of worms, tiny soil animals, and microorganisms, whose activities are essential to keep my plants healthy. Compost also supplies plant nutrients, making the soil more fertile. (This last function could be performed by fertilizer instead.) Since neither my own compost nor the feedlot manure I purchase supply enough plant nutrients to grow a really abundant garden, I also feed my veggies COF.

This approach seems inexpensive and practical *at this time*. Each year my garden goes through about three 50-pound (25-kilogram) sacks of seedmeal, one sack of meatmeal, one of kelpmeal (and the price of that always makes me wince), one or two hefty little sacks of high-phosphate Peruvian guano, about 25 pounds (10 kilograms) of ordinary agricultural lime, the same amount of gypsum, and 50 pounds of dolomitic lime. In 2005, the cost of all that came in under $300. This expenditure grew about $4,000 worth of veggies if their value is calculated according to what I would pay for them at the supermarket at the exact time I harvest them, which is an entirely unfair comparison. Still, spending $300 to produce an underpriced $4,000 seems a great exchange.

So each year I cheerfully pay out that $300 to maximize the nutritional qualities (and output) of my garden. But if my local farm supply store could not obtain the ingredients for COF, or if I could not afford them, I am glad to know I could grow almost as good a garden, producing nearly as nutritious an outcome, by putting out the effort to make high-quality compost and buying only a couple of bags of lime that would, at current prices, cost me about $20.

Why do I say my veggies are "underpriced" at $4,000? Because I am not growing stale, flavorless vegetables. And no allowance is made for the superior nutritional qualities of my food. Over the last two decades, according to official US government statistics, the average nutritional content of industrial vegetables has declined about 30 percent. And 20 years ago supermarket veggies still weren't the equal of home-garden stuff. I have no doubt that what I grow is at least twice as nutritious as the stuff in the supermarket and is easily worth $8,000.

That figure is still an understatement. Because I have been well-nourished for several decades, I need little medical attention, even though my body has arrived at its mid-60s. Entirely disregarding the dollar cost of sickness, I do not disregard the cost of the suffering. Add that to the value of my garden's output! ■

Why compost?

Nature recycles. Everything that grows will eventually die, fall to earth, and rot or will be eaten by large animals whose manure falls to earth and rots. Eventually the animals themselves fall to earth, and their bodies rot. The only example of a natural process I can think of that doesn't seem like recycling is when dissolved mineral nutrients end up in the ocean, which seems to be a huge permanent sink for nutrients. But even oceanic minerals recycle over geologic time.

Nature has designed wild plants to be capable of growing lustily in soil whose rather low fertility is stably maintained by organic materials rotting on its surface. I discussed in Chapter 2 how vegetables, lacking the vigor of wild plants, have come to require higher levels of nutrients in their soil and, often, higher than natural levels of organic matter. Early farming in all English-speaking countries consisted of clearing the forest and then turning under the duff, a thick, half-rotted layer of leaves, bark, and other organic matter. The duff is the forest's capital accumulation of centuries. Dug in, it decomposes far more rapidly, resulting in a huge but temporary increase in fertility. Abundant harvests could be enjoyed — for a while. But without the annual nutrient addition from the forest's falling leaves, with the export of soil nutrients as crops were sent to market, and without the addition of the manure produced by all the animals (including the humans) eating that food, soil fertility decreased, and all too soon the land was "worn out."

In this industrial era, exhausted land has been temporarily restored to heavy production by the use of chemical fertilizers. Were it not for the use of fertilizers, most of our croplands would be considered "worn out," and almost any soil a vegetable gardener uses these days may be too infertile to grow most kinds of vegetables. So increasing soil fertility is the gardener's main concern.

Mulch gardening

Since nature maintains fertility through the slow surface decomposition of organic materials and does not dig (except if you count the slow activities of worms), why not copy it: spread organic matter on top of the soil and let it rot. Will doing that grow a veggie garden? The answer is a qualified yes. It will grow veggies in climates where the soil freezes solid in winter *and* where summers are hot enough to rot the mulch rapidly. To get the best possible result you have to mulch with something rich in plant nutrients like alfalfa (lucerne) hay or pea straw, or else fortify infertile spoiled hay and autumn leaves with strong manure or heavy sprinklings of the same seedmeals that make up COF.

Here are some more negatives.

- The mulch is always in your way. It tangles the hoe but fails to completely prevent weeds from emerging. If you're mulching on the cheap, with spoiled hay, then you are simultaneously sowing huge quantities of grass seed. You'll end up spending more time patrolling for weeds than you would do if you'd not mulched and learned instead how to properly sharpen a hoe.

- Mulching keeps the surface moist but does not, as claimed by mulching enthusiasts, significantly reduce moisture loss. Soil moisture is mainly lost as plants transpire water, not as the sun shines on bare earth.
- Mulched gardens are slow to warm up in the spring.
- With permanent mulching, so much leaf and straw residue is being brought into the garden that the soil's nutrient ratios are inevitably put far out of balance. Consequently, the nutritional content of your veggies is hugely degraded.
- If your garden is larger than a postage stamp in the middle of a big lawn that easily supplies heaps of grass clippings, you will have to haul in a great deal of bulk to maintain the mulch. You'll do so much hauling, in fact, that it almost becomes a necessity to own, fuel, licence, maintain, and pay the insurance on a pickup truck. If you don't own one, then someone willing to haul for you has to own and use one on your behalf.
- And in mild-winter climates, after a year under permanent mulch the garden becomes home to plague levels of small animals that proceed to eat most kinds of vegetables, especially seedlings. I tried mulching in two different mild climates, which is why I do not recommend this method where the soil doesn't freeze solid in winter.

It works out to be less effort and far more effective to heap-compost enough organic material to maintain soil humus at healthy levels and then use something like COF or poultry manure compost as fertilizer, when needed, on medium- and high-demand vegetables. Mulch gardening does make sense for someone who is weak or whose physical mobility is restricted, someone like 70-year-old Ruth Stout, the author of *Gardening Without Work,* which started the whole mulching mania, someone too frail to wield a shovel or hoe but who can manage to carry a flake of hay to throw atop a weed coming through and who can afford to get someone to haul in and stack multiple truckloads of spoiled hay bales for her.

Mulching? I've been there, done that! If you want to reinvent the wheel, well, a lot of gardeners do. I reinvented quite a few wheels myself.

Carbon to nitrogen ratios

You can predict how any particular material will behave when being composted, and how well it can fertilize plants after decomposition, by knowing how much carbon (carbohydrate) it contains in relation to the amount of nitrogen (protein) it contains (see Figure 7.1). This is called its carbon-nitrogen ratio, sometimes written C/N or C:N. Not only can every decomposable material be compared with every other by its C/N, but there also is an absolute standard of comparison — the C/N of soil.

Except in the driest hottest deserts, where the earth contains no organic matter, all soil contains humus, the stable residue of decomposed organic matter. Humus is a complex substance highly resistant to further decomposition. It does eventually break down completely and vanish from the soil, but that happens very slowly.

Soil humus has a carbon-nitrogen ratio of around 12:1 in every climate, in every soil. Amend soil with something decomposable and the amendment will be converted into a far smaller quantity of humus. If the original C/N of that material is higher than 12:1, the soil microbial population will "burn" its

Carbon-nitrogen ratios

± 6:1	± 12:1	± 25:1	± 50:1	+ 100:1
Bonemeal	Vegetables	Summer grass	Cornstalks (dry)	Sawdust (500:1)
Meat scraps	Garden weeds	Seaweed	Straw (cereal)	Paper (175:1)
Fish waste	Alfalfa hay	Legume hulls	Grass hay (poor)	Tree bark
Rabbit manure	Horse manure *	Fruit waste	Cardboard (corrugated) †	Tree needles (conifer)
Chicken manure	Sewage sludge	Grass hay (green)	Tree leaves (deciduous)	
Pig manure	Silage		Autumn grass	
Seedmeal	Cow manure *			
Tankage	Spring grass			
Hair/feathers	Garden soil			
	Comfrey leaves			
	Coffee grounds			

* Containing no bedding

† The glues used contain a lot of nitrogen, lowering the C/N compared to paper.

Figure 7.1

carbon for fuel while preserving the nitrogen until the C/N lowers to match that of the surrounding soil. At the end of this burning off there will probably be a higher level of humus in the soil then there had been before the soil was amended. In simple terms, the soil becomes healthier and will have better tilth. Also, once the average C/N (soil plus amendment) comes close to 12:1, the plant nutrients that had been contained in the undecomposed organic matter become available to plants growing in the earth; again in simple terms, the soil becomes more fertile. Why did the soil not become more fertile immediately? Because until all the amendment is eaten, the soil microbes assimilate the nutrients and withhold them from the plants.

If soil is amended with organic material having a C/N lower than 12:1, the opposite happens. To increase the C/N to match that of the surrounding soil, different sorts of bacteria will convert the surplus nitrogen to ammonia gas. This reduction of nitrogen content continues until the stable desired ratio of humus, 12:1, is reached. The ammonia is almost instantly converted by other soil-dwelling bacteria into water-soluble nitrates, first-class fertilizer that makes plants grow fast. The resulting higher soil nitrate level also encourages soil microbes to multiply. More of them means they more aggressively attack humus. The end result is a soil with a slightly lower quantity of organic matter than there was before this low C/N material was mixed in. This loss of humus is also what happens when chemical fertilizers are added to soil, which is why organicists are so strongly opposed to the use of chemicals.

This deserves to be said again. If we put something into the soil with a C/N below 12:1, whether it is chemical or "organic," the consequence is the immediate formation of nitrates. The plants rapidly start growing faster, but when it is all over we have slightly reduced the soil's humus content, meaning a less-healthy microbial population and poorer tilth. If we decompose high C/N organic matter in soil, the result is that soil microbes co-opt nitrates and other plant nutrients present in the soil in order to eat that organic matter. For a time the plants grow less well. But when things settle back down, we have increased the soil's humus level and its fertility. And after the main decomposition, the plants grow better than before.

Finally, the higher the C/N of the amendment, the longer it takes to decompose; the longer the soil chemistry is disordered, the longer we have to wait before our vegetables grow well. That wait can be ruinous. Many gardens have been wrecked for an entire growing season or even longer because they were heavily amended with high C/N materials.

Sheet composting

Here's another good idea: Fertilize by spreading a layer of nutrient-rich, decomposable, organic matter atop the soil and then shallowly digging it in. Things rot much faster when they're mixed into well-oxygenated soil than when they're merely spread on the surface. This method, called "sheet composting," will grow a good vegetable garden — if you do it far enough in advance that there is time for it to decompose before planting time. And if you are sheet-composting fresh animal manure, it only works well as fertilizer if you spread *fresh* manure and, in hot sunny weather, incorporate it *immediately* after spreading it. In a personal e-mail, a Texas farming consultant I am acquainted with said of this:

> Throw manure on the top of the ground on a cool day without wind and the values might hold for several days. Put the same manure on the ground in 98 degree sunshine on a windy day and you will lose 25 percent of the nutritional values by sundown. The best with manure is to put it under the soil as quickly as possible — the shortest dwell time between the cow and the soil means the highest efficiency of conversion.

Farmers raising livestock frequently sheet-compost. They haul a few days' accumulation from the barn or loafing area to a field, spread it over a small part of that land, and immediately disc or rototill it in. Over the course of a month or so, they fertilize an entire field.

In the North American garden, you can bring in a load of fresh manure just after the first frost, spread it, and shallowly dig it in. At this cool season there wouldn't be a huge loss of nitrogen if it took a few days to finish turning it under. You can sheet-compost with other things than manure, but it is mighty difficult to spade in grass hay; alfalfameal is diggable but mighty expensive. Short grass clippings and leaves can be dug in. But if you're thinking of digging in high C/N vegetation, keep in mind that the length of time needed to decompose it depends on the following things:

- The thickness of the layer
- The average C/N of the material at the start of the process
- The preexisting fertility of the soil
- The amount of time the soil will be reasonably warm before you need to plant it.

Immediately after the blanket of decomposable material is dug in, the soil ecology begins to consume it. Microorganisms multiply incredibly rapidly on this new food supply. If the soil is warm, in a few days they'll be going at the material in much the same way yeast fills a brewing vat. These microbes obtain the raw materials to construct their bodies by gobbling up almost all the available mineral nutrients in the soil around them, nutrients that plants would otherwise use to grow with. The microbes especially need nitrogen compounds to form proteins, but they also need all the other usual plant nutrients: phosphorus, potassium, calcium, magnesium, etc. In other words, when undecomposed organic matter is mixed into soil, even though that organic matter contains a lot of plant nutrients that will become available after decomposition, these nutrients are not soluble, are potential, are not available yet. But the organic matter creates a microbial bloom whose need for nutrients is so great that plants growing there are deprived of nutrients.

Strange, isn't it! You sheet-compost to fertilize, but for a time the soil becomes less fertile. As the microorganism population busily "burns" the carbon, they also incorporate plant nutrients contained in the organic matter into their bodies. Microorganisms are more aggressive than plants in this respect. They get what they need first. Only when this new food supply has been almost entirely consumed does the population of decomposers begin to die for lack of fuel. At this point the plant nutrients held in bacterial bodies are released back into the soil for the plants to use. Some of these nutrients will be simple water-soluble chemicals, much like those that are applied as fertilizer. But some of them will be in the form of complex organic chelates that nourish the plants much as vitamins nourish humans. To be fully healthy, to achieve the highest nutritional quality for the humans using them as food, vegetables need to assimilate a goodly portion of their nutrients from decomposing microorganisms.

Like every other organic chemical or enzymatic reaction, the speed of sheet composting depends on soil temperature. In moderate or short-season climates it goes slowly in autumn, does not happen at all during winter, resumes in spring, and by the time the land has warmed up enough to start heat-loving crops, the decomposition process will almost certainly have been completed — unless the material being sheet-composted started with an extremely high C/N. In early spring, because massive decomposition is still going on, chill-tolerant spring crops may not grow well, so in regions with freezing winters, avoid autumnal sheet composting on beds where spring crops will be going in unless you're using low C/N material — and not too much of it. Where the soil does not freeze or chill too severely during winter, sheet composting goes on much more rapidly, unless there is sawdust (incredibly high C/N) mixed into the manure.

You can sheet-compost in spring or summer if you can afford to have some land out of production during the prime growing season. In warm soil it usually takes around six weeks for decomposition to proceed to the point where vegetables get the benefits of all the nutrients added. That's six weeks if the average starting C/N of what was turned in was no higher than 30:1 and the layer being amended isn't over an inch (2.5 centimeters) thick.

A few more cautions:

- Gardeners in well-rained-on regions who depend on sheet composting as their main or only method of soil improvement should spread lime along with their compost material. Fifty pounds (25 kilograms) of ordinary agricultural lime per 1,000 square feet (100 square meters) will not cause overliming. It's even better to spread 50 pounds of an equal mixture of agricultural lime and dolomitic lime per 1,000 square feet. Add lime at that rate every year, or at least every time you sheet-compost. However, beware if the manure is not from your own animals: it may have already been mixed with quite a bit of lime for odor control, and the result may be too much lime for safety.
- The soil in new garden sites will not contain enough nutrients to build the enormous microbial population needed to rapidly decompose a lot of material. If you're sheet-composting to start a new garden, don't spread low-potency horse/cow/sheep manure more than an inch (2.5 centimeters) thick. Avoid having much bedding material in the manure (Figure 7.1 explains why). Don't sheet-compost a thick layer of autumn leaves and

woody grasses of late summer on a brand-new piece of infertile ground. These materials also have a high C/N (consult the table). Do lime. Calcium and magnesium are just as vital to the formation of microbial protein as they are to making plant protein. I wouldn't like to have my readers discover in spring that their sheet composting had not finished and that nothing would grow well until midsummer. This is a common catastrophe.

- It is not wise to sheet-compost with poultry or rabbit manure; these strong, low C/N materials are best used as fertilizers. Plants will grow better immediately after low C/N materials are mixed into soil — if you don't mix in too much and end up poisoning the plants. However, if you are sheet-composting leaves and dried vegetation or old dried manure, and the average C/N of this stuff is high, the average can be lowered a lot (and the whole composting speeded up considerably) by including a thin layer of poultry or rabbit manure. It is the average C/N that determines the outcome.

Sheet composting is risky. Before you incorporate organic matter into soil, it is safer, and the outcome more certain, if you decompose it first so its C/N is lowered until it approaches 12:1. This is done with heap composting, which will be explained later in this chapter.

Temperature and decomposition

I have already mentioned that the speed at which organic chemical and enzymatic reactions go on is determined by temperature. To repeat: The rule is that for every 10°F (5°C) increase in temperature, the speed of the reaction doubles. This increase is geometric (with each 10° F increase, the speed goes from 2 to 4, 8, 16, 32, 64, 128, 256), not linear (the speed goes from 2 to 4, 6, 8, 10, 12). It is not quite that simple, however, because organic chemicals and enzymes are not able to withstand high temperatures. Enzymes are usually destroyed at around 120°F (50°C); a few, such as the enzyme that converts brewer's malt into a sugary wort or the enzymes that can work in a steaming hot compost heap, can handle temperatures as high as 155°F (68°C). If it's hotter than that, the enzymatic reaction instantly and permanently ceases.

The temperature at the soil's surface under a thick mulch during summer is cooler than the air temperature. In summer in the northern United States, for example, mulched soil might be no more than 75°F (24°C) averaged over 24 hours. Decomposition is slow at that temperature, so the release of nutrients in the mulch is slow. A six-inch-thick (15-centimeter) layer of spoiled hay or grain straw might take a year to decompose at the average year-round temperature of temperate climates, both because of low average temperatures and also because only the material in direct contact with the soil is decomposing at any significant speed. There will not be many plant nutrients being released.

But suppose you took the same amount of hay before it spoiled, passed it through a cow's gut, and then spread the cow's manure (and urine) and sheet-composted it. All that material is now in close contact with the soil. It was also thoroughly mixed with the cow's digestive enzymes and inoculated with all sorts of intestinal microoganisms. In this case, the process of decomposition is pretty much completed in only a few months, even in the chilly soil of spring. The nutrients are released much faster and the soil becomes much more fertile.

What is this thing called compost?

Composting decomposes organic materials before they are put into the soil so they have become instant plant food. Compost contains large quantities of recently deceased soil microbes, whose nutrient-rich bodies are virtually water soluble. It resembles loose, dark brown soil. With luck, it is mostly humus. This sort of compost only forms during an intentional, controlled, microbial ferment, much the same process used to make beer, wine, yogurt, sauerkraut, vinegar, aged cheeses, or tempeh. Food fermentations are delicious, but only when made with precision. Ferments can also be done casually, with highly erratic results. It is the same with compost.

For the ultimate in "erratic," take my grandmother Anna's Rosh Ha'shanna wine. Grandma Anna always made a batch of kosher grape wine for the High Holy Days. A month before the big event she would buy a few pounds of fresh purple Concord grapes and crush them between her hands into a big mixing bowl. Then everything — skins, seeds, twigs, juice, bowl — was covered with a piece of cheesecloth to keep the fruit flies out. The bowl sat on the kitchen counter. It was too early in the year to start firing the coal furnace, so unless the oven was going, the temperature in the kitchen was pretty much the same as whatever the temperature outside was that year. Wild yeast arrived on the skins of the grapes themselves, so, of course, the yeast was different every year. The sugar content of the grapes fluctuated wildly from year to year because sometimes they were purchased before they were fully ripe (the date of Rosh Ha'shanna was calculated on the Jewish lunar calendar, so it could be any time from mid-September to mid-October). If the juice seemed a bit tart, Anna might add a cup of white sugar. And of course nothing was sterilized before she started, so all sorts of other microorganisms contributed to the fermentation, including the ones that make vinegar. The mash would bubble away for a few weeks, and then Anna would pour it through a strainer. With most of the sediment removed, she would pour the raw wine through a few layers of cheesecloth to remove the fine particles, then pour it into an old whiskey bottle. If this wine sat around for only a week or two before the big Holy Day meal, it wouldn't be *too* vinegary when we sipped it in tiny glasses. You may be sure that no one drank more than the required minimum to sanctify the event.

Serious winemakers would never do as my grandmother did. They precisely regulate every aspect of the process — the temperature of the fermentation, the exact sugar content, the sterility of every utensil. They would carefully select a pure strain of yeast to suit the type of wine being made. And despite all those controls, the result still varies, which is why we have gourmet magazines and wine experts.

People who make homebrewed beer know exactly what I mean. Some homebrew is really foul. Oh, if you raise the octane high enough with table sugar it'll get you drunk, but it'll have a sour aftertaste, like hard cider does when it is made with baking yeast, and if you drink more than one bottle you'll have a splitting head the next morning. But homebrew made skilfully with the best ingredients, then aged for a few months on the shelf before you pop the cap, can be so superior to most industrial beer that you can hardly stand to buy the ordinary stuff.

Let me underscore the two essential differences between a talented vintner's wine or a proper homebrew and what my grandma Anna did. (1) Complete control of the conditions. (2) Standardization of materials — regulation of sugar content, the correct pure strain of yeast to inoculate the ferment. The same is true of making compost. If you avoid a few big mistakes while tossing almost any material that is handy into a heap, get it damp, and let it work, you'll end up with compost of a sort. But if you want high-quality compost that grows magnificent things, you have to be actively and intelligently in charge of every part of the process.

This reality has been hidden from home gardeners to a degree. That's because the organic farming and gardening movement was fomenting a social revolution when it first started in the 1940s. For propaganda purposes it was made to seem that all compost was good compost, and that any compost would

Analyses of various composts

Source	N%	P%	K%	N+P+K%	Ca%	C/N Ratio
Vegetable trimmings and paper	1.57	0.40	0.40	2.37	?	24:1
Municipal refuse 1	0.97	0.16	0.21	1.34	?	24:1
Municipal refuse 2	0.91	0.22	0.91	2.04	1.91	36:1
Municipal refuse 3	1.1	0.6	0.9	2.6	?	?
Missouri tree service (wood waste)	0.4	0.06	0.2	0.6	1.13	88:1
Missouri farm	1.8	1.5	1.4	4.7	2.35	24:1
Missouri home garden	2.1	0.1	0.5	2.7	0.16	21:1
Gainesville, FL, refuse	0.57	0.26	0.22	1.05	1.88	?
Garden compost A	1.40	0.30	0.40	2.1	?	25:1
Garden compost B	3.50	1.00	2.00	6.50	?	10:1
CA commercial for veg A	1.6	?	?	4.2	?	15:1
CA commercial for veg B	2.0	?	?	6.8	?	13:1
Poultry manure compost	3.9	?	?	7.6	?	7:1
Poultry manure compost	3.6	2.5	1.7	7.8	6.94	7:1
Poultry manure compost	4.0	2.6	3.9	10.5	7.14	7:1

Note: These figures have been gathered from various sources over many years; not all sources provided data for all columns.

The column "N+P+K%" is a useful gauge of overall mineralization.

Figure 7.2

do a great job at growing food. The lie was told so many times that those telling it came to believe it themselves. But if compost is going to noticeably improve the growth of most kinds of vegetable crops, it has to have a nitrogen content in excess of 1.5 percent and a C/N no higher than 15:1, and it must contain significant amounts of phosphorus, potassium, calcium, magnesium, and trace minerals, all in the right proportions. Not all composts do; most, in fact, don't, as shown in Figure 7.2.

Please notice in the table that municipal refuse compost isn't potent enough for vegetable growing. That's because much of the material going into the heap is paper and cardboard, chipped tree trimmings, and other woody waste. All these materials have a high C/N, and the average C/N of the heap when composting starts exceeds 50:1. After much heating and watering and turning and shrinking and rewatering and turning some more, the C/N drops to about 25:1. In other words, far more than half the carbon has disappeared. Why "far more" and not exactly half the carbon as those numbers might suggest? Because the way municipal compost is made causes most of the nitrogen to disappear as well. The final weight of compost will be only a tiny fraction of the starting weight. Thus municipal composting, portrayed as socially positive, actually is a huge contributor to greenhouse gasses.

Municipal composting operations find it impossible to get market gardeners to accept their product. Ultimately its destiny is to be mulch along roadsides, under ornamentals in parks, and sometimes in homeowners' yards and gardens. Those who foolishly try to use it to grow vegetables are sorely disappointed.

What I would classify as low-grade compost, barely suitable for a garden, contains around 1.5 percent nitrogen, its carbon-nitrogen ratio is around 20:1, and its total of N+P+K is under 3 percent. In climates where the soil gets warm if you shallowly rake in a small amount (maybe a layer a quarter to a half inch thick — 6 to 12 millimeters), this low-grade stuff will barely feed the sort of vegetables I call "low demand" (see the sidebar in Chapter 2).

To modestly grow healthy medium-demand vegetables in most climates and soils, what I would classify as medium-quality compost needs to be added to the soil at about the same rate noted in the previous paragraph. Medium-quality compost contains more like 2 percent nitrogen, has a N+P+K of around 4+ percent, and has a C/N no higher than 15:1.

To grow high-demand vegetables in almost any soil and climate, using compost as the sole source of fertility, the high-quality compost you'd need would have a nitrogen content exceeding 3 percent, an N+P+K exceeding 6 percent (with a fair amount of phosphorus in that 6 percent), and a C/N no higher than 12:1.

Did you notice that I kept qualifying my statements in the three preceding paragraphs by saying "most soils and climates"? That's because, as with every other chemical reaction, the release of nutrients from compost depends on the soil's temperature. Some places, particularly maritime climates, have rather cool summers; their soils never do get warm. There, you need to use somewhat larger quantities of more potent compost to provide the high levels of nutrients required by high-demand vegetables. If you are compost-gardening in a cool climate, the spring and early summer garden tends to get going rather slowly because the soil hasn't warmed up enough to begin cooking the nutrients out of decaying organic matter. In situations like this, you will do better using concentrated nutrient sources like COF in addition to amending compost or well-decomposed manure if you want to enjoy a good result with high-demand vegetables.

In hot climates organic matter rots quickly. In really warm soil it is possible to get an acceptable result growing high-demand vegetables with only medium-quality compost.

Making low-grade compost

It is difficult for a vegetable gardener to avoid making a compost heap. Composting was once prohibited in many cities because it can breed vermin when done carelessly, but now municipal governments encourage it. If the goal is not high-quality compost, the practice is thrifty and almost foolproof. Almost no matter what a gardener does or doesn't do, a heap of plant waste will eventually recycle itself into something resembling soil. Even a huge heap of sawdust beside a defunct sawmill will decompose — given about half a century.

Composting in bins

Mainly for aesthetic reasons, most city-dwelling and suburban homeowners compost in some kind of container. There are various prefabricated designs;

many are simply an open-bottomed plastic cylinder, four to five feet (120 to 150 centimeters) across, at least three feet (90 centimeters) high, sometimes bigger, and usually with a vermin-proof lid. The best of these have an opening at the bottom large enough to insert a shovel so the process can go on continuously. Lawn and garden waste and kitchen garbage are tossed on top as they appear. Sometimes water is added if the contents get dry. The container heats up, a little or a lot depending on what is put in, but it never heats all the way through unless the whole thing is filled all at once with a uniformly low C/N mixture. The contents usually shrink and settle rapidly enough that there is room on top for the next batch of yard waste. From time to time a bit of well-rotted material can be shovelled out of the access port at the bottom. If it is done right, the box keeps rats and mice out and there is next to no offensive odor except, perhaps, for the few moments the cover is removed to add new material.

Is the product of one of these bins effective enough to grow a great veggie garden all by itself? Rarely. But it will grow nice flowers and shrubs. Usually too much woody, decomposition-resistant stuff goes in. Even if the user has sense to avoid putting in woody wastes, the whole heap doesn't heat uniformly. There's usually a temporary hot spot just below where the newer materials are added. This area then cools too quickly to ferment properly. It is awkward to turn the material in order to loosen it and allow in sufficient air. The final result lacks potency and may lack mineral balance. Gardeners making this kind of compost and using it in the veggie garden will do far better if they also use COF or sacked chicken manure, as described in Chapter 2.

Tumblers

Another device sold to conscientious city and suburb gardeners is the compost tumbler. You're not told in the propaganda for these expensive gadgets that the drum may rust out. But the most negative thing about compost tumblers is promoted as their most positive feature — their speed. The idea is that every time you visit the garden you (almost effortlessly, so they say) crank the handle half a dozen times, which rotates the barrel once around. One rotation daily loosens, lightens, mixes, aerates. This makes the decomposition process go quickly, taking just a few weeks instead of months. All that seems a good idea ... when you read the advertisement. The tumbler also makes compost in small batches, which is how most gardeners accumulate compostable materials. So what am I grumbling about?

Research has shown conclusively that the more frequently a compost heap is turned (beyond the minimum necessary), the lower its value as fertilizer. Until I receive different data, I will consider a tumbler to be a convenient, high-speed recycling machine that replicates the municipal composting yard, not a producer of the finest garden fertilizer. And should you compute the ethics of recycling with one, then add to the cost of buying a tumbler the pollution and nonrenewability of the resources used to make the short-lived thing.

Grinders and fast compost

When I was a novice gardener, I read and believed glowing reports about UC Fast Compost in the Rodale Company's magazine. I purchased what I divined (from the magazine ads) was the best shredder-grinder. It wasn't cheap. It never occurred to me at the time that maybe the reason the magazine was promoting fast composting was because it had so many advertisers selling grinders. Maybe you, reading the next paragraphs, won't buy one. As I said earlier, my father told me the cheapest experience people can get comes secondhand — if they'll buy it.

The idea behind grinding is that the smaller the size of particles in the heap, the faster it'll decompose and the hotter it'll get — if turned frequently. And it's a lot easier turning a heap of fine stuff than it is to fork over a heap of stringy material. However, one of my Internet acquaintances, who once ran a commercial composting operation, told me, "Research has shown conclusively that compost heaps that get hotter than about 150°F [65°C] lose heaps of valuable nitrates and burn out too much carbon. If the material has been cured at about 155°F [68°C] for several weeks it off-gases methane, ammonia, and carbon dioxide. These gases contain the bulk of the nutrients. What is left is little more than ash. Think about putting what you eat for lunch into an oven for the same time period at the same temperature as occurs with this kind of compost pile — what you get will be ash." Imagine how efficiently you could dispose of organic wastes by combining a grinder with a tumbler!

Home-garden-sized grinders need hefty, fuel-hogging engines. Grinding is hard, brutal work for both machine and operator. The machines need a lot of maintenance and parts replacement. I never found grinding to be pleasant: the operator needs to wear goggles, gloves, and ear protectors, and may breathe a lot of dust. Instead, why not take it easy? Save the gasoline. Don't contribute to the unnecessary manufacture of short-lived machinery. Let the microbes in an ordinary heap do all the work for you at a slower pace, at a temperature that doesn't ruin the value of your product.

Medium-quality compost: The once-a-year heap

It is far better to compost with the intention of making something I would call "medium-quality." If you have no livestock (except perhaps a few chickens), and if your garden is under 3,000 or 4,000 square feet (280 to 370 square meters), you'll have better results if you build only one large compost heap a year. In temperate climates, the most convenient season for this is early autumn, just after the summer crops have been cleaned up and the deciduous trees (if you have them) have dropped their leaves. (If you have livestock and a larger garden, or a big lawn contributing a lot of grass clippings, it might be best to start a new heap every time you've accumulated sufficient material.) (However, you'll have a far healthier lawn if you allow the clippings to rot in place instead of removing them.)

Here's what I do. Starting in late autumn, I steadily accumulate all new vegetative wastes and kitchen garbage into one great stack of plant material that I encourage to dry out over the following eleven months. If anything is more than a foot (30 centimeters) long, I use a sharp machete to chop it up into smaller bits while it is still green and tender. This is because long stringy pieces will later be painfully awkward to turn over. This material, topped off by a large addition from the autumn garden cleanup, is what gets composted. I am a country gardener with no neighbors complaining about my untidy heap of dry vegetation, so I can spread all the new stuff on top of the vegetation pile in the order that it appears — pea vines; the contents of the kitchen compost bucket; trimmings, like outer leaves of lettuce or carrot and beet tops, that appear when I harvest; old cabbage and cauliflower plants; weeds; and straw from producing my own vegetable seeds. This material rapidly dries out in the sun.

City or suburban gardeners, or those with a greater need for tidiness than I have, might wish to make some sort of corral for these materials. These folks probably will also be adding grass clippings. (I have no lawn; my house is surrounded by native vegetation kept roughly mowed by wild grazing animals.) If you are accumulating grass clippings, it is far better to allow them to dry for 24 hours before you rake them up and spread them thinly atop the pile to fully dry out. By making "hay" in this way, you prevent the clippings from heating up (and losing nitrogen) before they dry out. .

After the first frost comes, and after the autumn garden cleanup, I have built a large heap of well-mixed dry material, the result of months of layering fresh vegetation and kitchen garbage. When I make the compost heap, I attack the dry material stack from one end, uniformly blending everything. This is important because when this stuff is made into compost, I want the average C/N of the initial heap to be fairly uniform and, I hope, below 25:1.

To keep the average C/N down, I put in nothing woody, nothing with even thin bark on it: no tree trimmings, no hedge trimmings. Trust me on this; you don't want wood in your compost! The C/N of woody materials is way too high and will degrade your compost quality. For the same reason, don't put in any sawdust if you want your end product to have much fertilizing power. If you seem to accumulate a lot of woody plant residues, make a separate compost heap of them outside the vegetable garden. Follow the same directions I am about to give you for composting the better-quality, lower C/N, dry vegetative wastes. The wood compost pile will be a slow-working, low-temperature heap, and it might be several years before it seems done. When it does seem finished, do not put it in the vegetable patch. Spread it as mulch under ornamentals or under fruit trees.

Some of the worst pests overwinter on or in the dead vegetation of your garden. If you till it in before autumn, you reduce their population, but you also leave the earth bare over winter. This is not a good practice! Burning the dead vegetation is also a poor practice because it destroys organic matter that the soil needs. However, in the high heat of a properly made compost heap, these insects will be destroyed. This is one more reason to compost — and to compost properly and thoroughly.

Composting procedure

In autumn, my year's accumulation of dried garden vegetation and kitchen garbage becomes a compost heap. The heap is built right beside the dried vegetation. I need the following additional material:

Water. I need enough water to thoroughly moisten the entire mass of dry vegetation. The easiest way to get it is to have a hose and nozzle handy. If running water is not available, about 100 gallons (400 liters) of water should be enough to moisten a year's accumulation from a garden of about 2,000 square feet (200 square meters). If you lack running water, and spreading a barrel of water with a watering can is not to your taste, you could spread the dry material out like a mulch and wait until it gets well rained on, then gather it back up and meld the damp vegetation into a composting heap.

Soil. Soil in a compost heap fills the same function as moderator rods in a nuclear reactor. It slows the process down a bit and captures the radiation — not of neutrons, as in the case of a reactor, but of nitrates. When the heap starts working, proteins in the decomposing vegetation will be broken down into ammonia, a gas. Without soil close by, ammonia will escape into the atmosphere and be lost, to the enormous detriment of your compost. However, soil commonly contains a type of bacteria that rapidly converts ammonia into nitrates, which are not gasses. *Putting enough soil in the heap to support these bacteria is more important than putting in manure or other strong sources of nitrogen.* When composting a heap composed mostly of vegetation, use enough soil so that it makes up a generous 5 percent by volume of the initial heap. (Two wheelbarrow-loads of soil are sufficient to moderate my garden's annual vegetation collection.) Too much soil is far better than too little. It will be best if you use fertile soil from your garden, and it's especially wise to take some of it from the spot where the previous year's compost heap was cooked. This way the heap is also inoculated with appropriate microorganisms.

Strong stuff. You'll want something with a C/N much lower than that of the dry vegetation. And enough of this strong stuff that it will lower the average C/N of the entire heap well below 25:1.

The finest thing to use for this purpose is truly fresh ruminant manure without any bedding. Horse and cow manure is usually not high in nitrogen, so a large quantity of it is needed to bring the heap's average C/N low enough, which means that including manure of this sort will significantly increase the amount of finished compost you get. If it is fresh, it also contains large quantities of still-active digestive enzymes that help speed the initial breakdown of the vegetation. How much should you use? That depends on how much bedding is mixed into the manure. If there is little or no bedding, you could add a volume about a third as much as the heap of dry vegetation. If bedding makes up more than half the manure's volume, there is no quantity that will be large enough. Manure plus bedding in equal quantities should be composted by itself with soil; its own average C/N probably exceeds 25:1. If manure plus much bedding is mixed into a pile of dry garden vegetation, you'll need to incorporate some other highly potent material to lower the average C/N enough.

The next best material is dried poultry manure without bedding. If you have a chicken coop, you should collect and store their droppings in used feed sacks. Accumulate this dry manure for the annual compost heap. How much should you mix in? Assuming that the average C/N of the vegetation heap, before any chicken manure is put in, is around 30:1, then an addition of about 5 to 10 percent by volume of the entire heap would be about right.

Least best is seedmeal, such as I suggest for making COF. Seedmeal is about twice the potency of chicken manure. I need a single 50-pound (25-kilogram) sack for my year's accumulation of dry vegetation. This is what I use because these days I have no chickens.

Once I have the materials, I take the following steps:

Building it. Starting at one end of the heap of dry vegetation, use a pitchfork (and/or your hands) to pull vegetation out of the end of the pile of dry stuff and spread it in a layer on the ground. Make a rectangle at least five to seven feet (150 to 210 centimeters) across, at least five feet long, and about eight inches (20 centimeters) thick. The heap needs oxygen to work, and if you build it too wide or too high, the core will be relatively airless, so make it no wider than seven feet, but as long as you like. Cover this rectangle with about a half-inch-thick (1.25-centimeter) layer of soil. Cover that with about three quarts (three liters) of seedmeal or two gallons (8 liters) of chicken manure or an inch-thick (2.5-centimeters) layer of horse or cow manure. Then water the whole mass long enough that the vegetation on the bottom is thoroughly damp.

Now repeat the process. Spread another eight-inch-thick layer of dry stuff, then more soil, then more strong stuff, then water. Repeat this layering again and again until the entire pile of dry vegetation has been layered into a new heap. What you are building will inevitably taper as it goes up. If you estimated width and length correctly, then the height when you are finished will exceed four feet (120 centimeters). It's better if it is five feet (150 centimeters) high, but should not exceed six feet (180 centimeters).

Spread a thin layer of soil over the entire outside of this heap. Don't attempt to thickly frost the heap with so much soil that you cover every last bit sticking out, but this final layer of soil will greatly help the outside of the heap to decompose more effectively and will also capture ammonia that might otherwise escape.

Heating it. Within two days the pile should be heating up. In the chill of early morning you should see some steam coming out. You may have read that heaps must be very hot to be effective. It's not true. You'll end up with more potent compost if the core temperatures are only 125°F to 135°F (50°C to 57°C), although if you've erred on the side of making the average C/N too low, the temperature can exceed 155°F (68°C). If you've not lowered the C/N enough, the heap will not heat or will heat only slightly. To check this temperature, push your hand deep into the heap. It should be uncomfortably hot but not so intense that you can't tolerate if for five or ten seconds. If you made it get too hot, you're going to end up with less volume than you could have achieved and it may not be quite as potent as it might have been; vow to do better next year. If it isn't heating much, it would be a good idea, when you turn the heap the first time, to blend in some more strong stuff or some stronger stuff than you used the first time.

Cooking it. Within a few weeks the fermenting heap should sag quite a bit. The fermentation should be so vigorous that even if there are autumn rains, the heap's core will dry out. Also, as it sags, the core temperature should start dropping.

The first turn. If you live where winter is severe, turn the heap before there is any chance that things will freeze solid. No matter where you live, certainly turn it by the time five weeks have passed. Next to the heap there should be a bare area where the original stack of dry vegetation sat. Using a hay fork and/or shovel, move the pile onto that spot. Start by peeling off the outsides. These should end up in the core of the new heap because these materials will not have decomposed much. The core of the existing heap got hot; it will have decomposed the most. This original core material should end up toward the outside of the new heap.

While turning it, remoisten the material. Get it good and damp again, but don't make it so soggy that if you squeeze some in your fist, water squirts out.

The new heap will be smaller. Make its outside dimensions so that it is at least four feet (120 centimeters) high. When it heats back up, it won't get as hot as it did the first time, nor will it stay hot quite as long.

Subsequent turns. If you garden where there is a hard winter, the heap will probably freeze and stop working shortly after the first turn. When it thaws out, wait a month, turn it again, and moisten it if necessary. The heap should promptly heat back up. Wait another month and see if it's done. If not, give it one more turn.

I live where winter consists of some hard frosts; some cold, windy, rainy days; and some lovely, sunny, autumn-like days. In my climate I give the heap a second turn about the time the apple trees are blossoming. Some years it does not need a third turn.

I think the difference between needing two or needing three turns depends on how much strong stuff and how much soil I put in. More of either results in fewer turns and a quicker process. Each turn is easier to accomplish than the previous one because the material has broken down more, so you can do more with a shovel and less with a fork ... and the heap has shrunk. My compost is ready for use before hot-weather crops like tomatoes are put in.

Doneness. A compost heap that did not contain bark, twigs, or sawdust will be done enough to use when it is dark brown in color, is crumbly and loose, has a pleasant odor, and contains nothing that resembles the stuff it was originally made from. It will probably contain some worms.

A compost heap that contained bark, twigs, or sawdust might be cool and look done after three turns, but it will still have a high C/N and might not grow vegetables well until another entire year (and several more turns) has passed. There will be little volume left after that, and almost all the value will have off-gassed (see Figure 7.2).

The garden is not a closed system

The waste from my own garden, roughly twenty 100-square-foot (10-square-meter) vegetable beds, plus the kitchen compost bucket will build an annual heap of loose, dry material about six feet (180 centimeters) wide by seven feet (210 centimeters) long at the base and roughly five feet (150 centimeters) high. That includes the autumn cleanup. Late the next spring, when this material has fully composted into a much smaller volume, I discover that I have made only enough to cover a third of my beds about a quarter inch (six millimeters) deep. But to maintain the proper level of soil organic matter, all my beds should be amended each year with compost or well-decomposed manure about a quarter inch deep.

I apologize again for being the one to give you more bad news, but no vegetable garden can operate as a closed system that, all by itself, sustainably generates enough organic matter to maintain itself *unless it is cover-cropped at every opportunity and unless every bit of the organic material it generates is recycled — and that includes your own humanure.* A farm sagely run on sound biological principles and endowed with reasonably fertile soil can slowly increase its overall level of soil organic matter and can even export a fraction of the organic matter it produces (in other words, the crops sold to market) without loss of fertility. But a garden? Never!

If you recycle everything from the garden (except humanure) back into it via the compost heap, and if, in addition, you import lawn clippings and a large autumn accumulation of deciduous leaves (neither of which I have) into the garden's system, would that be enough organic matter and fertility? It depends on how much material you have and how rich it is. For me, in my climate, on my soil, with my lifestyle, I must import and compost a heap of dry organic matter at least twice the volume of my own garden's output. That import could be ruminant (horse/cow/sheep) manure, spoiled hay, autumn leaves, or used coffee grounds from the espresso shop. Whatever it is, it should have an average C/N of about 25:1, more or less the same as the annual heap of dry vegetation I save up from my gardening and compost.

What I actually do to eliminate this deficit is use the services of a neighbor who makes his living hauling spent brewer's malt from the local brewery to a feedlot about 30 miles (50 kilometers) from my home. Instead of coming home empty, he fills his large dump truck with well-rotted feedlot manure and offloads the manure at my garden gate for a small sum plus the cost of the material itself, which is not high at all. One load tops up my garden's organic-matter deficit for about two years.

Someone with an extra piece of deep, well-drained soil about the same size as their vegetable garden could grow a comfrey patch. The enormous output of mineral-rich biomass from the patch can make the entire household operation become a closed biological system. Comfrey is an extraordinarily

To make the best comfrey tea, loosely pack a barrel three-quarters full of freshly cut leaves and then fill it to the brim with water. Allow to steep for a week or two. Comfrey tea is often sprayed on plants as a foliar fertilizer and general tonic to prevent disease. Filter it first. In either case, whether you're using it in a spray or for fertigation, dilute it half and half with water. After steeping, drain the barrel. The solids may be dug into the earth or composted. ■

deeply rooting and aggressively growing plant whose leaves can be cut repeatedly during spring and summer. The leaves may be fed to livestock (after they have wilted into "hay"), included in compost heaps, and used to make effective fertigation tea. Once it's established, you can keep the patch growing vigorously by running chickens in it, so long as the poultry are also fed grain. The comfrey patch might also be the ideal place for sheet-composting humanure.

Constraints on the final size of my book do not give me the space to discuss the fine points of establishing and managing a comfrey patch, but if you have some extra land and are interested in maximum self-sufficiency, I suggest a respectful reading of *Russian Comfrey: A Hundred Tons an Acre of Stock Feed or Compost for Farm, Garden or Small-holding* by Lawrence Hills (see the Bibliography).

Composting manure

If you own livestock, you'll need more information than this book could contain. However, if you know people who own livestock and are foolish enough to sell or give away their animals' manure, I can set you up to handle it effectively.

One thing you *do not* want to do with any raw manure is incorporate it in your garden just before starting a vegetable crop whose edible parts will have contact with the soil. Some manures are contaminated with antibiotics, hormones, disease organisms, and pesticides. Composting will go a long way toward eliminating these. Certified organic growers are not allowed to use uncomposted manures on any vegetable crop unless more than 120 days have passed between incorporation of the manure into soil and harvest. Raw manures decomposing in soil can adversely affect the flavor of vegetables. Raw manure is best used on cover crops (see the section later in this chapter) or sheet-composted immediately after the growing season to overwinter and fertilize the next year's growing effort.

Ruminant manure compost. Manure (except for poultry manure) comes in two parts: solid and liquid. They are equally valuable. Often the urine goes to waste, which is most unfortunate. If the animals are penned or stabled and bedded on straw that soaks up much of their urine, be sure to compost this too. If the urine has soaked into sawdust, forget about using it to make a compost that'll grow veggies unless you live where the soil gets mighty hot in summer.

The C/N of the solid fraction of animal waste varies, depending on what the animals are fed. Horse/cow/sheep solids could run anywhere from 10:1 to 15:1. It could contain a lot of valuable minerals in addition to nitrogen and potassium, or it might not. Urine-soaked grain straw might have a C/N of 30:1. Urine/straw and solids mixed together in more or less equal quantities provide the basis for a fine compost pile, so long as enough soil is added to make up 5 to 10 percent of the total starting volume.

If there is no bedding straw, the next best thing is to mix the solids with a more or less equal quantity of dried vegetation, such as chopped corn stalks, spoiled hay, leaves, garden waste, etc. It is probably okay to use up to half again more volume of dry matter than manure in the heap. Don't forget the soil, and don't forget to adjust the moisture content so it is nicely damp.

Compost and turn as usual. If you start out with a heap whose C/N is close to 30:1, and if you do your part turning and keeping it moist, then you can expect the resulting compost to adequately grow medium-demand vegetables.

If you have neither vegetation nor bedding straw to mix into the solid manure, mix it with soil. Use about a quarter soil by starting volume. It will still heat quite a bit. If it gets too hot to insert your hand and/or if it starts smelling like ammonia, then break down the heap and mix in more soil. Turn this kind of heap every two weeks. It won't take long to compost. You'll know it's done when it stops heating up more than a little bit, even though it's still moist. Because its ingredients started out with a much lower C/N, this compost will probably end up being more potent than the stuff made with vegetation or bedding, and it may be good enough to grow high-demand vegetables. (In the old market-gardening books, this was called well-aged "short manure," because "long" manure had the straw.)

Poultry or rabbit manure compost. Poultry have no urine to be wasted, so their manure is naturally more complete. Poultry are usually fed grain, and productive free-ranging chickens usually supplement their grain with food that contains even higher amounts of protein — tender green grass and insects, for example. As a result, their manure is highly potent.

Rabbits are almost always caged and fed mainly on pellets made of alfalfa and grain, though sometimes they are given garden trimmings. Their droppings fall below the cages and are mixed with their urine. Rabbit manure is also potent stuff. If you keep rabbits, you can improve the compost you'll make by spreading a thick-enough layer of grain straw under the cages to soak up the urine. When you start smelling ammonia around the cages, it is time to rake up the whole accumulation, compost it with soil, and spread new straw to start the next batch.

Compost poultry and rabbit manure just as you would other forms of manure, except for two things. The C/N of the manure is well under 10:1, so you'll want to prevent huge losses of nitrogen by mixing it with enough vegetation to bring the starting C/N of the heap to about 20:1. Also, for the same reason, it's even more essential to include soil in the heap when you're composting strong manure.

Do beware of allowing the heap to overheat. If that happens you'll be losing a lot of valuable nitrogen. Any time you can't insert your hand into the heap and keep it there for a short time, the heap must be cooled down. If you have a long probe thermometer, make sure the heap doesn't exceed 140°F (60°C) at the core. The best way to moderate a too-hot reaction is to turn the heap and mix more soil into it as you turn it.

Properly composted poultry or rabbit manure will end up with a C/N of close to 10:1. It will contain at least 3 percent nitrogen. Poultry manure will also contain lots of phosphorus (the seeds chickens eat are phosphorus rich). This stuff will grow high-demand vegetables.

However, be leery of commercially prepared chicken manure compost or pelletized chicken manure products. Usually these are made by blending manure with sawdust. After a brief initial heating, the material cools and a laboratory analysis will show a high NPK, but the sawdust's decomposition process will have barely begun. The material is then dried and sacked or pelletized. However, when you put it in the soil, some of the nitrates in the chicken manure are locked up as the soil life decomposes the sawdust. The result obtained with nitrogen-demanding vegetables is not what you might expect.

It will work okay, but only if you use larger quantities than would be indicated by the lab analysis. Measured by result for dollar spent, you will often have better results using seedmeal or other organic concentrates like feathermeal, fishmeal, or tankage. And of course I am convinced that the best result of all comes from using COF.

Making the highest-quality compost

I have not mastered the art of making the highest-quality compost. I have read about it at length. I have attempted it and never quite succeeded. My conclusion is that a home gardener cannot realistically aspire to making the finest possible compost because to succeed, one must use large quantities of the freshest manure (before any nitrate losses occur) of the most potent sorts. Therefore, the composter must own a barn full of livestock or a large coop of chickens and must also institute management systems that effectively capture all the urine. I well remember the ammonia smell that developed under my rabbit cages when I raised rabbits. This odor proved that the manure I had accumulated had already degraded. To make really first-class compost out of bunny manure, you'd have to process it about once a week, before any smell developed. And to do that and come up with enough material to make a heap of minimum size, you would have to have a great many rabbits and be operating on a scale far greater than the average homesteader would want. You would be in the rabbit business, and you'd be making a new compost heap each week, year-round. So you'd have to be selling compost as well as rabbits. And you'd have to have plenty of dry vegetation, with a not-too-high C/N, to mix into the brew. Every time I consider the subject, I conclude that making high-quality compost is only for the small farmer who possesses a great deal of energy.

The most illuminating information I ever read about composting was in one of the earliest books written on the subject: *The Waste Products of Agriculture*, by Sir Albert Howard, the founder of the organic farming movement. It is cited in the Bibliography and is easily obtainable as an e-book if you have online access. It is also an interesting read that illuminates conditions in British-ruled India during the early 1930s.

The biodynamic agriculturalists also know a great deal about composting. There are books by Koepf and Pfeiffer listed in the Bibliography, and I recommend them highly.

Humanure

It has never pleased me to write book reports, even of great books. So let me simply say of Joseph Jenkins' *The Humanure Handbook* that it fully explores the possibilities and techniques of recycling human waste. All the data in this book is the product of a family's actual longstanding practice. *The Humanure Handbook* can be purchased in print through the usual channels of retail trade or downloaded at no cost from many websites. If you'll consider the subject without preconception or revulsion, you'll see that, other than the relative inconvenience when compared to the prevailing system, there is no reason not to compost your own family's humanure and use it to grow vegetables. However, to avoid having your currently healthy family pick up humanure-transmitted diseases (mainly parasitic ones), it might be a good idea to segregate the humanure of visitors and houseguests.

My wife, Muriel, and I have thoroughly discussed restoring this type of waste management to our own household. Restore? Muriel was a Tasmanian child in the early 1940s. She grew up in a low-income family on a self-sufficient, rural, half-acre homestead. The family's meager water supply came from small tanks that captured rainwater off their roof. It is a small joke between us that in Tasmanian folk culture, all vegetable gardens were fertilized exclusively with animal manure — except in those places that had no flush toilets. Muriel's family garden depended on humanure, as did the gardens of most of her neighbors. Our chats have led to a thorough look at the chores and methods needed to prevent the "dunny" (the bucket containing humanure, or the outhouse in which the bucket was placed) from odorizing the entire place. Muriel insists that a crudely built and extremely well-ventilated outdoor cubicle is the best kind of dunny, meaning one goes outside to poop (though the house has chamber pots for use at night, during stormy weather, etc.). As we do not live where the soil freezes solid for months on end, it seems that to try recycling humanure where the snow flies for months on end would be particularly daunting. However, the author of *The Humanure Handbook* lives in Pennsylvania, where there is some real winter, and he says otherwise.

If times got tough enough that I could not afford to spend a few hundred dollars each year on maintaining the garden's fertility, or if the materials to do so got so scarce that there was little choice, then we would definitely start using humanure. We owe thanks to Joseph Jenkins for having so bravely opened this discussion at a time when those in industrialized countries may really need to master the recycling of this valuable resource.

Green manure and cover crops

People visiting my veggie garden in midsummer will marvel at a dinner-plate-sized head of broccoli from a three-foot-tall (90-centimeter) by four-foot-diameter (120-centimeter) plant, or at the vibrancy of some other crop that seems to exceed their skill. They will decide this proves what a skilled grower I must be. But these same visitors usually miss the even greater significance of a few thick patches of waist-high buckwheat scattered here and there. To me these bits of cover cropping offer far better proof of what they imagine me to be. Green manuring/cover cropping is a skill I'd like you to have, too. But practicing it can be tricky and a bit hazardous.

For those who are unfamiliar with the words, green manures and cover crops are interchangeable terms for growing something whose sole purpose is enriching the soil. On a farm scale, a cover crop is plowed in or knocked down and allowed to rot on the soil's surface. On a garden scale, it may be mowed, the tops hauled to the compost heap, the stubble and root systems dug in. In the garden, a few cover-cropping species can be conveniently yanked out of the ground by hand, en masse, and the whole plant composted.

Why do it?

If a plot of land is put to work growing as much biomass as can be formed during every day of the year that biomass will grow, and if almost all that biomass is recycled back into the land, then that bit of earth will arrive at a relatively stable and high organic matter content and will maintain it indefinitely. In that condition of maximum organic matter content, the soil will have good tilth and grow good crops (unless it is an infertile bit of earth).

An example of what I am describing is a pasture growing a carefully selected mixture of grasses and clovers. There is a thick sward of grasses catching every bit of sunlight and converting it to biomass every possible moment. No hay is removed and no animals graze there. Suppose that land were merely mowed once or twice a year and the clippings allowed to rot in place. Over a few years of such treatment, the fertility (soil organic matter level) would build up to a climax level and stay there indefinitely.

An extreme opposite would be a vegetable garden. In the garden, just at the time of year when the plants would be growing their fastest, we till the land bare and plant little seedlings or sow seeds in rows far apart. By the time our vegetables' leaves again thickly cover the ground (have formed what is termed a "crop canopy"), much of the growing season has already passed. Usually the crop comes to maturity and dies back before the growing season is over. Considerable growth could happen between the time of the first frost and the hard freezing of winter, but the veggie garden, which grows mainly frost-tender species, is bare then, too.

Cover cropping means making sure the ground produces more biomass by covering it with a crop canopy for as much of the growing season as possible.

The downside

How-to books extolling cover cropping rarely indicate the risks. So let me start by listing the negatives.

Green-manuring practices have filtered down to vegetable growers from farmers. We gardeners use shovels and sometimes light powered equipment. Farmers use big tractors. Before tractors, they used horses and oxen, sometimes several at once ganged to pull heavy plows. Farmers using (horse)power could grow cover crops that were mechanically tough, but the home gardener can be overwhelmed trying to handle a six-foot-tall (180-centimeter) stand of woody-stalked grain rye. I know. I have grown small-scale cereal rye by hand. My point? If we gardeners are going to green-manure, we had best use types of crops that match our strength. So beware of advice from holistic farming books. And doubly beware of blindly following advice found in holistic gardening books and in government agricultural extension or ministry of agriculture publications aimed at the home gardener. A lot of this information was cribbed from half-understood farming books but was never practiced by the author; many young extension agents are faced with publishing or perishing, but haven't more than a taste of practical hands-on experience.

Using the wrong species of green-manure plant, or green-manuring on clayey soils, combined with a bit of bad luck in spring, can ruin all chances of having a timely planting. In early spring, all overwintering green-manure crops are tender, their vegetation contains a great deal of nitrogen (in other words, their C/N is low), and they will decompose rapidly in soil. However, if you let the crop continue to grow until it starts forming seeds (or until flowers appear), the vegetation rapidly becomes woody and the C/N increases markedly. This is significant! When you dig tender, low C/N, green material shallowly into warm earth, where there is lots of moisture present, it will decompose almost completely, roots and all, in two weeks. However, if you give that vegetation a few more growing weeks, time to get woody, and dig it in — if your shovel and strength *can* still dig it in — it will now take four or five weeks to decompose enough to allow seeds to sprout. Why do I say "allow"? Because (as discussed earlier in this chapter) when a great deal of decomposition is going on in the soil, and when the microbial population has enormously increased (that's what decomposition is), then these microcritters are breathing in heaps of oxygen and exhaling heaps of carbon dioxide gas, and the soil atmosphere becomes so filled with CO^2 that seeds cannot even germinate. At the same time, if transplanted seedlings don't die outright because their roots are suffocated by too much CO^2, they will not be allowed to grow. All this settles back to normal as soon as most of the decomposition has finished.

To summarize: If you turn in tender, high-nitrogen material, you have a two-week wait. If you allow it to get what the farmers call "forward," you are looking at a four- to five-week wait before you can plant.

Why do I make such an agonizingly detailed point of this? Well, imagine having heavy soil and a thriving overwintered green crop, say a stand of rye grass or winter wheat coming out of the snow. Spring weather goes against you and it rains more than you'd expect. When the time comes to dig in the knee-high grass, the soil is too wet to work. If you go ahead and dig, you'll make such a mess of clods that no seedbed will be possible, maybe not all summer. If you wait for the land to dry out, the grass will be waist high (if it's rye, it may be chest high) and forming seeds before the earth has dried enough to permit tillage. So the weather has already made you late to prepare the soil, and now you'll have an additional four or five weeks to wait after you dig in the cover crop. That land might not be ready to plant until it is time to start your cool-weather crops for autumn.

Wintering over a cover crop is much safer on sandy soils that won't form clods if you have to dig them in when the soil is wet. But it's not that much safer. Light soils can quickly dry out in a rainless spring. Dig a cover crop into dry soil and it won't decompose. Now how do you work up a seedbed? And when rain comes and it does start to decompose, nothing will grow well for a few weeks.

It is much safer for gardeners in areas with harsh winters to use autumn-sown cover crops that will certainly be killed by winter. Their roots will decompose in early spring. Then in spring, when you can dig the garden, there is a much higher chance of ending up with fine, loose seedbeds.

If you are gardening in semi-arid country and don't have much irrigation, and if it turns out to be a relatively dry winter followed by a rainless spring, the green-manure crop might suck so much moisture out of the soil before you turn it in that the ground will be deprived of the moisture you need to get the garden going.

One last warning: Beware of a phenomenon called "alleopathy." The term means that one crop puts residues into the soil that harm the growth of the crop that follows after it. There is a possibility that the green manure/cover crop you select could prevent some of your garden crops from growing well, which is the opposite of the effect you hoped to achieve.

The upsides

And what are the positives of green manures/cover cropping if all goes well?

- The soil is thickly filled with rapidly decomposing roots, leaving the land loose, airy, and totally open to rapid penetration by your crops.
- Beneficial insects have had a place to overwinter.
- If the green manure was a legume, it will have put some nitrogen and organic matter into the soil. If it wasn't a legume, it still put organic matter into the soil.
- As an autumn or overwintering cover crop grew, it incorporated a great many plant nutrients that might otherwise have leached out of your garden during heavy rains. Now these nutrients are part of the decomposing biomass, to be slowly and steadily released to your growing crops.

Specific suggestions

The best book I ever read about this practice is *Using Cover Crops Profitably* (see the Bibliography). It was written by a committee of American horticulturalists because every district, every climate, needs to use different species, on a different schedule. One person could not possibly grasp and convey all that detail with any real experience. Each kind of cover crop has a different freeze-out point, which limits its range. In one place it dies in autumn; in another it reliably carries over into spring. And in North America, as one goes poleward there is less and less time to establish a crop before the soil freezes.

Cover cropping is something that gardeners must work out for themselves — with guidance in the beginning. It takes a few years to develop an intuitive sense of how to use the crops that will work for you. The practice is of great interest to horticultural advisers; your local agency will have a good deal to tell you about what farmers in your area do. Ask them. A mail-order vegetable-garden seed company is not the most affordable source for cover-cropping seeds and may not sell the species best suited to your area; your local farm supplier is probably the best source.

Summer green manures. Buckwheat is the only summer cover crop I can recommend without hesitation, knowing it will work almost anywhere and not get any gardener into trouble. It grows best during early summer, and there are local varieties adapted to your situation. In the northern hemisphere, buckwheat sown thickly before the solstice will grow waist high in less than five weeks and then begin blooming. Seedmaking puts an end to vegetative growth. At that point it should be dug in (or yanked from the ground and dried for later composting). The stalks are both tender and brittle, so they are easily dug in or rototilled without tangling in a light tiller's tines. They decompose in just one week if the crop is dug into moist soil before more than the first few flowers have appeared. A buckwheat cover crop puts the soil into magnificent, fine-textured condition, a perfect seedbed for any following crop. Any time you harvest an early spring sowing (peas, spring salads, radishes, mustards) before midsummer and there is nothing scheduled to fill that space for at least five weeks, scatter the fast-germinating seeds so they are about one inch (2.5 centimeters) apart and chop them in shallowly with a hoe.

One warning: Buckwheat's flowering seems to be determined by photoperiod, so it is important to use seed that was grown in more or less the same latitude you live in. What do I mean? Once I used seed from the bulk bin at the local health-food shop. I live at 41°S latitude. The seed had been produced in semi-tropical Queensland, grown during the "winter" cool season at 25°S latitude. The consequence was that it hardly got to a foot (30 centimeters) tall in my garden before it went into full bloom and stopped growing. This may not be such a problem for North Americans.

In hot climates, southern peas will function in much the same way as buckwheat does. They have the additional benefit of being a legume, which means there will be nitrate formation. Promptly till in or yank out the crop when seed formation begins.

In the spring I have planted whole beds with ordinary garden peas as a green-manure crop. This practice has the benefit of filling our freezer with peas. If the beds are needed sooner than the peas mature, the green manure can be yanked and composted anytime. Peas leave the soil in the most beautiful condition!

Crude brassicas. Brassicas like fodder kale, oilseed radish, tyfon (a turnip x Chinese cabbage hybrid), rape, and field turnips make excellent autumn green manures. In cold climates they will certainly freeze out, not posing any problem in spring. Where winters are milder, they may overwinter. In maritime climates and other areas with mild winters they will almost certainly overwinter and will make lush spring growth that blooms early. In those areas, don't overwinter brassicas on clayey soils unless you can yank them out by hand when blooming begins. Be sure to sow them at least 40 days before the end of their autumn growing season. Their leaves protect the ground in autumn; if they fail to survive winter, their residues still continue to protect the ground even though the plants are dead. Their huge taproots will really open up the soil.

Annual legumes. Annual legumes are the mainstay for overwintering in maritime climates and anywhere else they won't freeze out. Beware of using perennial legumes; these tend to grow more slowly their first year and may be hard to get rid of (this is particularly true of red clover). In Oregon, my favorite winter-hardy annual was crimson clover. Hairy vetch is a popular one in eastern North America. I've been experimenting with lupins, small-seeded fava beans, Persian clovers, and seridilla; all these are eminently suited to maritime climates, and they might suit the middle states in the United States (note that I said"might"). I know for a fact that crimson clover is widely grown over winter as far north as southern Kansas. Ask your local experts — governmental or else the feed and grain dealer — for the best prospects in your area.

Cereals and grasses. Avoid using cereals and grasses unless you own heavy power equipment and garden on light soil that dries out quickly in spring.

As I prepared to write this book, I discovered the new practice of no-till transplanting (of, say, a tomato crop) into a mulch made from overwintered grain rye grown right on the site (sometimes the rye is mixed with hairy vetch). Beware: I've never done it. Apparently the rye is cut off at ground level just after it has formed seed heads. The dead vegetation is allowed to remain lying in place as a thick mulch. A gardener prepared to mow (by hand with sickle or scythe) an incredibly thick six-foot-tall (180-centimeter) stand of rye might find this just the thing. It works because the decomposing rye secretes a chemical that blocks the sprouting of weed seeds. The process is not apple-pie simple, however. The decomposing vegetation will be rather woody, so unless some strong source of nitrates (COF or chicken manure, for example) is sprinkled on the earth before the rye is cut, its presence may tie up nutrients, resulting in retarded growth during the beginning of the season.

I suggest you Google "cover cropping vegetables" and see what turns up. Best you search this topic for yourself; I am reluctant to include URLs in a book because they change so rapidly. (I wonder. Will someone reading this book a century from now have even a clue about what I mean by "URL" and "Googling" something?)

CHAPTER 8

Insects and diseases

I doubt that in any single lifetime a human could experience the garden pests and diseases found in all climates. I intimately know only the ones in Cascadia and Tasmania. However, having spent decades dealing with these pests, I believe I can also make sensible suggestions about insect problems east of the Cascades and in the United Kingdom, even though I have never personally experienced them.

Avoiding trouble

Sir Albert Howard, founder of the organic farming movement, believed that before a plant is attacked, it has already become unhealthy. The plant predator's purpose in nature's scheme is to restore balance, like a wolf pack bringing down a sick, old animal that has lived too long. It has become organic-movement doctrine that a truly healthy plant will either be unassailable or will outgrow insect damage and will successfully resist disease. To the organicist, the key is making perfectly fertile soil and, thus, growing healthy plants.

I have noticed a lot of evidence to support this viewpoint. For example, in my variety trials

In my book *Growing Vegetables West of the Cascades* I covered one major (often disastrous) regional pest, the symphylan, that isn't addressed in the present book. Symphylans are only experienced as a problem west of the Cascades. I ask Cascadian gardeners to refer to the 4th or 5th edition of *Growing Vegetables West of the Cascades* (the earlier editions are badly flawed) for a level of specificity I can't match in a volume that is intended for gardeners in any temperate climate. ■

involving eight varieties of Brussels sprouts, all the plants of one poorly adapted variety will be seriously damaged by aphids, while all the plants of the other seven varieties remain almost entirely untouched. In another case, a disease will pick off a single tomato plant, while others of the same variety surrounding and even touching the diseased one will not be affected. This last event leaves me wondering if my distribution of soil amendments was not uniform or if that particular seed was somehow flawed.

Before I learned to directly seed whenever possible, I purchased many transplants. One spring after I'd filled my cabbage bed with garden-center seedlings, three seedlings remained unplanted. For an experiment, I put them into the unmanured fringe of the garden with only a bit of cheap chemical fertilizer for help — no compost, no COF, no lime. As the season went on, the difference was astonishing. Cabbages in the properly prepared bed grew big and healthy, with no problems. The extra three were attacked first by flea beetles and then by cabbageworms, both of which I sprayed. They grew so slowly that I side-dressed them with chicken manure, but the roots were severely attacked by maggots, all three became a bit wilty, and one died from this root loss. At summer's end the huge cabbages in the garden weighed six pounds (2.5 kilograms) each, and tasted fine; the two survivors on the fringe were one to two pounds (0.5 to 1 kilogram) at best, tough and bitter.

However, even when you provide ideal soil conditions, disease or predation may prevail when you encounter long-lasting poor weather or grow a variety or species not adapted to your climate or soil (like that Brussels sprout variety I mentioned). This can upset people with faith in organic gardening. Organicists also become dismayed when insect populations reach plague levels and the best-adapted variety, enjoying ideal soil and perfect weather, faces troubles that no vegetable can outgrow. An example of this is regularly experienced in the Skagit Valley north of Seattle, Washington, where thousands of acres are devoted to brassica seed crops. The poisons used to protect those plants kill off all natural predators. Consequently, cabbage pests that have developed resistances to pesticides become so numerous that local gardeners must also use effective controls or they cannot grow cole crops. (Note: These controls still do not need to be chemical poisons.)

You eastern American gardeners may consider yourselves fortunate to rarely meet the carrot rust fly or the cabbage root maggot. In Cascadia, fly maggots riddle carrots and leave them inedible; another kind of maggot destroys the root systems of more delicate brassicas and regularly ruins turnips and radish crops. In the fertile Willamette Valley, gardeners have major problems with these two pests; even more so do people in western Washington state. Yet when I gardened in Oregon I had no rust fly trouble and few cabbage root maggots because I lived in the Coast Ranges, worn-out land that was no longer farmed, where the sad-looking pastures were full of wild carrot (Queen Ann's lace), wild cabbage, and wild radish. Large, stable populations of hosts for these pests also meant large stable populations of their predators. There was plenty for the fly's larvae to eat outside my garden, and there were plenty of other living things to eat the fly. However, where there are garden carrots but few or no wild carrots, or garden cauliflower but no wild radish or wild cabbage, then the flies can quickly breed into a serious plague that goes entirely unchecked by predation.

My point is that urban gardeners and those living in prosperous agricultural districts are inevitably going to have more problems with pests. Keep this in mind if you're thinking of buying a new home(stead).

One way backyard gardeners in thickly settled territory can fight back is to increase the numbers of beneficial insects around their gardens by creating proper habitat for them. Someone with a bit of acreage can do a great deal to provide permanent cover that assists beneficials. This is a complex subject that I can't cover adequately in this book because every climatic zone requires encouraging specific plants that do not also aid insects you wish to discourage. Rex Dufour's excellent article "Farmscaping to Enhance Biological Control" is available free online from ATTRA (see the Bibliography) and can provide initial guidance.

On the other hand, there's almost nothing a gardener can do when weather becomes too unfavorable for a species to handle, opening the door to disease or insect attack. I remember well one lousy summer when there were too many cloudy, humid, cool days. My tomato vines steadily weakened, and finally, during the damp chill of what should have been high summer, late blight disease blackened all of Cascadia's tomatoes. Nearly every single tomato plant, from Canada to northern California, died within a few weeks. The only survivors had been grown in greenhouses or under the protective eaves of a sun-facing white-painted wall, where more suitable growing conditions gave them strength to resist the infection. It helps when you understand this, so that a simple shrug of acceptance is possible. And it is easier to accept weather-related losses if you grow a diverse garden that produces half again more than you'll need and contains crops that prefer both hotter and cooler weather, a garden that contains a core of really hardy vegetables almost certain to produce something in any circumstance.

Pesticide versus fertilizer

Before you rush to spray poisons, even natural pesticides; before you invest in row covers or rush out to purchase some natural predators; before fighting ... please consider this: Is the struggling plant simply not growing fast enough to overcome the problem? Unfavorable spring weather could be the cause, but sowing too early for the species was your choice, one that retarded growth and lowered the health of the plant. Often the best cure is not "killer" but liquid organic fertilizer — a foliar spray such as combined fish emulsion and liquid kelp (a triple whammy, with two fertilizers, one of which temporarily disguises the plant's odor from predators). Maybe a bit of spot fertigation will save the day. Maybe in a week or so growing conditions will improve, the soil will warm up, nutrient release will accelerate, and everything will come right.

Figure 8.1: *Spot fertigation.*

Also, it's wise in such circumstances to take out insurance: immediately sow again! And sow a bit more thickly this time. When weather conditions are unfavorable, if you start many more seeds than the final number of plants ultimately wanted, you can have a relatively benign attitude as insects and diseases help thin out the weaker seedlings. This later sowing may grow faster from the start and end up yielding sooner than the earlier one.

Speaking of a benign attitude, remember my comments in Chapter 4 about buying seedlings at the garden center? It's easy to be relaxed about abandoning some seedlings when the seeds only cost you a few cents each, but it's much harder to be dispassionate when seedlings that cost a few dollars apiece are being knocked over.

As a last resort, pesticides (preferably natural substances and not dangerous chemicals) may provide a short-term solution while you wait for spring weather to moderate and growth to resume. Besides, in spring the plants are tiny, so only a little dab of poison here and there is needed. And nothing is blooming; you won't risk killing pollinating bees.

Spun-fabric row covers provide most of the benefits of a cloche or mini-greenhouse without your having to erect any structure. They also provide protection against flying insect pests without your having to spray. Several brands are available; all are about five to six feet wide (150 to 180 centimeters) and may be offered in cut lengths starting at about 20 feet (six meters) up to industrial-agricultural rolls thousands of feet long designed to be reeled out behind a tractor. These spun fabrics are not completely transparent. The amount of light lost varies by make, as does their longevity and durability.

The fabric is spread over a growing row or bed and loosely anchored with a sprinkling of soil. As the plants grow, they lift the almost weightless fabric so no supporting structure is required. To keep insects out, the entire perimeter must be carefully anchored with soil, leaving no cracks or openings. Neat stuff!

One brand, Reemay, provides a few degrees of frost protection and considerably increases daytime temperatures, enhancing springtime growth. However, Reemay also decreases light transmission by 25 percent — not a desirable side effect. Other brands weigh less than Reemay, are less abrasive to plants when it gets windy, do not reduce light transmission as much, and do not cause any heat buildup in summer — but may not last as long.

On the downside for all these products, they are usually made with oil or natural gas as feedstocks, are expensive, and can't be counted on to last more than one season, even with gentle handling. ■

Weather and decreasing light levels at summer's end can also prompt troubles with diseases and insects on heat-loving crops. These are hardly worth fighting, since the plants' life cycle is virtually over anyway. But if you find yourself fighting for a crop from start to finish, either something is wrong with your soil management, or else the crop is one that you should not be attempting on your soil type or in your climate.

Certain vegetables are fussy about the type of soil they will and won't grow in. These species may express their difficulty by being susceptible to insects or disease. This is especially common with weakly rooting species intolerant of clay soils, such as globe artichoke, celery, celeriac, melons, and cauliflower. Brussels sprouts are the opposite; they don't like to grow in light soils. If they are planted in such ground, the sprouts tend to blow up and be loose, and the tall plants fall over.

Gardening aikido

My wife, Muriel, and I believe that when one of us annoys the other, this irritation, rather than being an excuse to throw blame around and have an enjoyable fight, is an opportunity for the annoyed person to clean up some of his or her own rough edges. After all, if I didn't have something in my own character similar to what is irritating me, I wouldn't be bothered much when I see this trait in another.

In much the same way, I try to see pests as guides to becoming a more skilful grower. One important step along this path is abandoning what self-help psychologists and some preachers call poverty consciousness. Instead of planting a garden from which you'll only harvest exactly what you want if everything goes well, plant twice as much as you need, so that pests and diseases could wipe out a third of the garden without stopping you from giving away buckets of food.

I will confess here that I cannot raise decent celery. I could claim that a stalk-invading disease always cripples my crop about when the plants have grown to a foot (30 centimeters) tall. More likely what is really happening is this: The top foot of my garden is a magnificent loam; under that is a fairly dense clay. Celery won't root in clay, period. But the root systems must eventually enter that clay if growth is to continue. The roots can't enter the clay, growth virtually ceases, health ceases, and a disease organism arrives to pick off the weak members of my herd. There is little or nothing I can do short of creating a deep bed of special soil or selling the property. Shrug! ■

Some years just *are* difficult years: the sun doesn't shine often enough, it rains too much or not enough, the tomatoes get blossom end rot, melons and cucumbers succumb to powdery mildew or beetles, eggplants won't set or get covered with red mites, peppers get some virus the extension office never saw before, corn is late and not as sweet as you remember it and the last sowing fails to ripen before frost comes, snap beans are covered with aphids, cabbages are covered with aphids, Brussels sprouts are covered with aphids, aphids are covered with aphids. But if you planted twice what you needed, there will still be enough.

I have pointed out several times that planting too early is the biggest single cause of trouble, but I need to say one more thing about this. I call your attention to the fact that growth rates accelerate hugely as the soil warms up. Sowing on the first possible day that the species could germinate or survive if transplanted, and sowing again two weeks later, will result in two crops with only a few days' difference in maturity. But the crop sowed two weeks later will have a lot less trouble. You will soon see from your own experience that I am speaking the truth here.

Another way to avoid battling is to reconsider the American Sanitary System (ASS), something Everybody Else believes in. Everybody Else's ASS suggests that food should be "clean" and free of bugs. The belief system is so widely held that no one in our supermarkets would buy spinach with a few holes in the leaves or a rutabaga (swede) with a few scars on the skin. Imagine the horror of finding a blanched cabbageworm in a frozen broccoli packet!

Why not change your attitude about what constitutes acceptable table fare? Remove as many bugs as possible when the food is washed, and then discretely slide the rare insect that escapes the cook's scrutiny to the side of your plate. Remember, there's a big difference between a plant showing the odd hole from an occasional insect and one that has been severely damaged. As long as the plant is still growing vigorously, a few (even a few hundred) pinholes in the leaves — or a scar on the cucumber's skin — don't matter. The critical level of leaf damage, where production and quality start to lessen, is a loss of about ten percent of leaf area. The critical level where aphid infestation has become too much is when they cover about 5 percent of the entire plant. As long as the cook can peel an occasional scar from the skin, why even bother to oppose the pest committing that minor nuisance? Let commercial farmers fight bugs on behalf of their unrealistic clientele.

Insects and their remedies

Aphids (*Aphadidae* family)

Often called plant lice, these are small, soft-bodied insects that cluster on leaves and stems, sucking plant sap. A few cause no significant damage unless they transmit a virus disease. In large numbers they cause leaves to curl and cup and can weaken or badly stunt a plant. Aphids can multiply with amazing rapidity, exploding from nothing to a serious threat in days. But they can also persist at a low level without causing much trouble. Don't rush to fight them the first time you see a few. To stop disease-carrying aphids from infecting your plants, you would have to spray something that would kill every single one before it entered your garden, and that's not possible! It's better simply to make your plants healthy enough to resist most diseases most of the time.

Aphids sometimes have a close relationship with ants, which "farm" them much as humans graze cows on pasture. Ants place aphids on leaves and then milk a sweet secretion from their livestock. Aphid control may entail elimination of ant nests.

You can spray aphids off leaves with a hose and nozzle. The ones you blast off will probably not find their way back, although others may. Safer's soap (a North American insecticidal soap made from special fats) is effective and virtually nontoxic to animals and most other insects, although it can burn the leaves of delicate plant species (I've especially noticed this with spinach) if used in strong concentrations. If you suspect you're going to have to use Safer's, test a bit on a single leaf a few days before you spray the whole patch. With Safer's, you also need to experiment with dilutions; it often works fine in greater dilutions than recommended, and when it's weaker it is easier on delicate plants. Australians have long used ordinary soap for this purpose — the traditional brand is Velvet, made from only tallow and lye. Americans have one like it called Ivory. The concentration required is about what you would use to wash dishes. When Muriel was a child-gardener, her family killed aphids by pouring the precious leftover dishwashing water over the plants. Preparations of rotenone plus pyrethrins appeal to people who enjoy killing bugs because just about anything hit with this combination drops dead almost instantly; both are natural poisons that (unfortunately) decompose in the environment within hours. This mix also kills bees and beneficials. The final remedy I recommend is neem spray, a long-lasting natural oil from a tropical seed.

Cabbage maggots *(Hylemya brassicae)*

See Root maggots, below.

Cabbageworms *(Pieris rapae) (Trichoplusia ni)*

There are at least two distinct species. Maybe four. The large green ones are larvae of a white butterfly often seen fluttering about the garden. Their clusters of small, yellowish, bullet-shaped eggs are laid on cabbage family plants, usually on the undersides of leaves.

A similar pest, sometimes called the cabbage looper, is a smaller larvae of a night-flying brown butterfly. Its round, greenish white eggs are laid singly on the upper surfaces of leaves. The larvae hatch out quickly and grow rapidly while feeding continuously on brassica leaves.

Both species can do a great deal of damage in a short time, especially if they begin feeding at critical times (such as during the early formation of cabbage heads) or if their numbers are excessive. Surprisingly, the small larvae of the night flyer are usually more destructive than the large ones of the day flyer.

In a small garden, handpicking and tossing the larvae away from any cabbage family plant can be sufficient control. An extremely effective nontoxic pesticide called *Bacillus thuringiensis* (Bt) is widely available, often marketed as Dipel. Bt can be sprayed the day of harvest because it is lethal only to the cabbageworm, cabbage looper, and a few close relatives. The culture remains active on the leaves for only a week or so, but even if sprayed only once, it seems to persist in the garden at a low level as infection is transmitted from decaying infected worms to healthy worms, significantly reducing their numbers for the rest of the season. If resprayed a few days after every rain or overhead irrigation, Bt can satisfy finicky growers who become grossly offended at the idea of a cabbageworm appearing in their broccoli or cauliflower heads.

Cabbage family plants usually have rather waxy leaves; sprays tend to bead up and run off of them. It is essential to put some sort of water softener (spreader/sticker) into the spray tank when using Bt. I use a quarter teaspoonful (1.25 milliliters) of cheap dishwashing liquid per quart (liter) of spray. It is also wise to spray the undersides of leaves as much as, or more than, the tops.

Carrot rust fly *(Psila rosae)*

See Root maggots, below.

Colorado potato beetle *(Leptinotarsa decemlineata)*

Found almost everywhere in North America, this beetle can almost completely defoliate a commercial potato crop. It may feed on other solanums (i.e., peppers, eggplants) and assorted weeds and flowers. The adults overwinter 12 to 18 inches (30 to 45 centimeters) below the surface in or close to the potato plot. Late in spring they emerge, lay eggs on the undersides of leaves, and resume feeding. Then their larvae also begin feeding.

If you grow spuds on a new piece of ground, few beetles will emerge immediately in the patch, which means row covers can be an effective defense.

Growing spuds on new ground is a good idea anyway, as it reduces disease problems, especially "scab." Straw mulching (using wheat or rye) after the seed has been planted has been proven to greatly reduce problems by providing habitat for beneficial predators that feed on both beetles and their larvae. However, this practice, which might work for the large-scale market grower, restricts your ability to hill up by hand hoeing (see Chapter 3), which helps prevent green potatoes and increases the home gardener's yield. What a conundrum!

In northern North America the beetles go through one egg-laying cycle each year; but in the southern United States they may go through as many as three. Getting the crop in early and growing early-maturing varieties may help southerners get the harvest in before the beetles get too thick. Rotenone/pyrethrum is effective for about two days and may be repeatedly sprayed, but it also kills beneficials — and in the long run, regular use of any poison works against the grower.

There are biological pesticides, but they have limited effectiveness. One is a strain of *Bacillus thuringiensis* that kills only potato beetle larvae and only when they are young. To make it work, you must watch closely for hatching eggs and then spray. There are more broadly effective fungal antagonists, and certain strains of parasitic nematodes have also been effective. Finally, for the home gardener who can spend more money than the entire crop is worth, there are flamers that lethally burn the beetles and their eggs off young plants.

The best and most affordable methods for the home gardener are to make sure there is lots of habitat around the veggie garden for beneficials and to spend a few hours handpicking adult beetles, especially attempting to eradicate them when they first emerge in spring, before they lay their eggs.

Corn earworm *(Helicoverpa zea)*

A close relative of the cabbageworms, this pest is also known as the tomato fruitworm because late in the season, when corn is not attractive to it, the insect may become a minor nuisance on tomatoes and, sometimes, snap beans. It usually does not survive freezing winter, so it only becomes a pest in the north of North America late in the season, after it has flown up from farther south. Food of choice for the larvae is ripening corn seeds; in California I have seen them eat more than the top half of an ear before it was ripe enough for me to try eating the bottom half. I've also seen a few in Tasmania, but not enough to bother fighting them.

The rather plain-looking small moth lays its eggs on the green silks of the corn plant. The larvae hatch out and follow the silks into the ear, fouling the ear with their excrement as they eat their way down the cob. Control is simple, effective, but a bit painstaking. Shortly after corn pollination is finished, mix a double-strength batch of Bt (with a spreader-sticker) and then, using a soft brush, generously daub this mixture on the silks of each ear. You could also make a blend of dormant oil (white oil), water, and Bt.

Cucumber beetles (Striped: *Acalymma Vattatum Trivittatum*) (Spotted: various species of *Diabroticae*)

Adults of all types of this North American pest are about a quarter inch (six millimeters) long. The striped ones have three parallel lines running the entire length of their back from head to tail; the *Diabroticae* have various patterns and colors of dots on their backs. Both species overwinter in the south and emerge in spring. In the north, the striped beetles overwinter while the spotted ones, strong flyers, migrate from the south, appearing about May/June. Winter/spring weather conditions can greatly alter their numbers. Some years are difficult for the gardener; some easy.

Overwintering beetles damage spring seedlings, chewing on leaves. Their larvae then proceed to feed on the roots, somewhat stunting the plant. The beetles prefer to feed on cucumbers. Their next favorite meal is cantaloupe and other similar melons, then squash, and watermelon last. There are also varietal differences in beetle interest.

If you have severe and repeated trouble, I suggest you do variety trials or consult your local agricultural agency about the best varieties for your area. Some years this damage can seem catastrophic. Many years it is minor. Emerging or migrating adults usually do minimal damage unless you are a market producer; they may scar skins of fruit or otherwise reduce the yield somewhat.

You can deal with this pest without directly fighting it. The easiest strategy is to be a few weeks later than most in your area to sow all cucurbits. This lets the seedlings get growing more vigorously, minimizing the significance of beetle damage. Another tactic is to plant four seeds, thinning to two plants per spot only after the seedlings have made a true leaf and are growing fast. (Where beetles are not a problem, one surviving plant per spot or hill is a far better practice.) Overwintering beetles sometimes transmit a virus wilt disease that kills seedlings. If a seedling does succumb to either predation or virus, there will still be growing time, in all but the shortest-season areas, to sow another in a different spot; the worst of the predation should have passed by then anyway. Excessive soil nutrients make seedlings more succulent and thus more attractive to beetles. And greenhouse-grown seedlings are always lush and succulent, having been pushed to achieve maximum growth rates. This case supports my contention that direct-seeding, a bit on the late side, is the best practice.

Putting each hill of emerging seedlings under a carefully anchored square of floating row cover will keep the beetles away until after they have laid their eggs. When the vines begin blooming you can remove the cover, allowing bees access to the flowers. Parasitic nematodes will control the larvae if you apply them to the root zone of small seedlings by mixing a dose of nematodes into a few quarts (liters) of water and fertigating the vine with them. However, these sorts of controls often cost more than they are worth.

Finally, the experts, each probably having read Everybody Else's manuals, recommend thorough cleanup of cucurbit residues, but I fail to see how this will do much good, considering the life cycle of the pest. If the larvae do mature, they'll overwinter under any available cover and won't need dead cucurbit vines for this purpose.

Flea beetles *(Phyllotreta striolata)*

These tiny, black, hopping insects chew pinholes in leaves, mainly in small members of the cabbage family, but occasionally they'll feed on other vegetables. In high numbers, flea beetles stunt and kill seedlings. Fast-growing, healthy plants usually aren't significantly perforated; tender greenhouse seedlings often have a hard time because they're in shock, not having been hardened off before being set out.

In early spring, overwintering adult beetles migrate into the garden from surrounding fields and begin feeding. Later in spring, the adults lay eggs in the soil. These eggs hatch into soil-dwelling larvae that feed on various roots, usually without doing much damage, until they pupate. After maturing into adults, the beetles then continue to feed until they hibernate in fall. This later feeding is usually of no consequence.

Flea beetles seem a pest only in spring when, due to cool conditions, plants are not growing rapidly and the garden is mostly bare, forcing many of them to concentrate on a few seedlings. Husky, well-hardened transplants normally outgrow flea beetle nibbling. Problems may be prevented by directly sowing five seeds for every plant wanted and then thinning only as competition starts to affect their growth. This gives the beetles more to chew on while providing the gardener with enough vigorously growing survivors to establish a stand. It is also wise not to sow at the earliest possible moment.

Keep a close eye on spring brassicas. If seedlings are losing more than 10 percent of their leaf area, the best first remedy may be a foliar feeding of fish emulsion (the foul smell of which might also confuse the beetles for a few days while the seedlings get going again).

If the loss of photosynthetic surface exceeds 20 percent, I suggest spraying every few days with rotenone or a liquid combination of rotenone and pyrethrum. Once weather moderates, the problem should go away.

Japanese beetle (*Popillia japonica* Newman)

Mostly located in the northeastern states of the United States, this pest is steadily spreading west and south. Its major economic damage so far is to turf, although it feeds on many kinds of plants. The beetle goes through one egg-laying cycle a year and spends perhaps ten months each year in the soil. The adults emerge in late spring and are gone within a few months, having laid their eggs.

You can spray adults feeding where they are unwanted with rotenone. However, they are unlikely to cause major damage in the veggie garden. If the larvae are feeding on vegetables' roots, it may not be easy to identify them as the culprits. There are strains of Bt and other biologicals specific to this pest that can be mixed in water and poured on the soil.

Japanese beetle traps can significantly reduce this pest's population levels if the entire neighborhood cooperates and everyone traps them. I stress "entire" because the adults are strong flyers.

Leafminers (*Liromyza spp.*)

See *Growing Vegetables West of the Cascades.*

Root maggots (*Hylemya brassicae, Pisla rosae*)

These larvae of innocent-looking small flies are only a major problem in Cascadia and the United Kingdom. I can't begin to tell you what a relief it is for a cabbage lover like me to live in Tasmania, a place where the weather is like Oregon's, but neither of these pests are present.

The cabbage fly waits until the root system of a brassica has become extensive enough to support its brood (this is about when the stem approaches a quarter inch — six millimeters — in diameter) before laying its eggs on the soil's surface near the plant. After hatching, the larvae burrow down and feed on the roots. Weaker-rooting brassicas — i.e., small-framed cabbage, most broccoli varieties, and all cauliflower, but rarely Brussels sprouts — become wilty. They may collapse and die or become stunted and barely grow. Maggots also tunnel through turnips, radishes, and the lower portions of Chinese cabbage leaves, although they tend to leave rutabagas (swedes) alone or at most scar up the thick skin, which is peeled away before cooking.

Some cole varieties have stronger root systems that tolerate a certain amount of predation without the plant wilting or becoming noticeably stunted. That's why I always did brassica trials without maggot protection. A magnificent result is when eight out of ten varieties are demolished by maggots and two varieties seem unscathed. In the case of radish, turnip, and Chinese cabbage, varietal choice provides no relief. Your only options are to use row covers or to undertake a timely harvest that gets them out of the ground before the maggots have invaded many roots. Unfortunately, "timely" harvesting is not an option with Chinese cabbage.

Gardeners can avoid much trouble by planting after the spring population peak. By early summer the spring maggots are pupating harmlessly in the soil. Mid- to late-May through July is the best time for North Americans to sow brassicas. Maggot levels increase again in late summer when the pupae hatch out, but by then non-root brassica crops are usually large enough to withstand considerable predation, and light intensity has dropped so much that even if plants do lose some root, they are not likely to wilt. Unprotected radishes, turnips, Chinese cabbage are still ruined.

In North America, this pest is a major problem only in Cascadia. Elsewhere it is a minor annoyance.

The late Blair Adams, research horticulturist at the Washington State University Extension Service in Puyallup, did extensive trials on a number of traditional organic remedies for root maggots. He found that dustings of wood ashes or lime — once widely recommended — actually attracted cabbage flies. He speculated these remedies helped despite that because in the (unlimed) acidic, calcium-deficient soils typical of Cascadia, the calcium-rich wood ashes boost the growth of brassicas enough to compensate for the increased predation the ash caused. Blair also found that careful and persistent hilling of soil around the plants' stems increased the survival rate of seedlings somewhat by burying the root system deeper.

The best simple organic control Blair could come up with was the collar. Gardeners had long used about a square foot (900 square centimeters) of tarpaper with a slit cut halfway through it so it could be fitted tightly around the stem, effectively keeping the fly away from the soil, but Blair felt that sawdust worked better and was easier to apply. A ring of fresh fine sawdust about 1½ inches (four centimeters) thick, six to eight inches (15 to 20 centimeters) in diameter, touching the stem, will prevent the fly from laying its eggs on the soil's surface. To protect radish and turnip, sow the seeds on the soil's surface, cover the seeds with a band of fine sawdust four to six inches (10 to 15 centimeters) wide and one inch (2.5 centimeters) deep. Timely harvest is still essential because the swelling roots push the sawdust aside and expose themselves to the fly's rapidly hatching eggs.

Since Blair did his work, another organic remedy has become available. Certain species of parasitic nematodes effectively attack root maggots in the ground. If large numbers of these microscopic life forms are seeded into the soil surrounding brassica seedlings, they can live for months, breeding and maintaining fairly effective population levels for a while, and actively knocking off maggots as fast as they hatch out. Parasitic nematodes will also control numerous other pests including wireworms, onion maggots, carrot weevils, cutworms, rhododendron root weevil larvae, strawberry root weevil larvae, and cucumber beetle larvae.

Parasitic nematodes are easy and cheap to culture by the billions, but it's not always so simple to transport or store them alive once they're out of the culture medium. Be cautious if you're buying them and make sure what you're getting is fresh and remains effective.

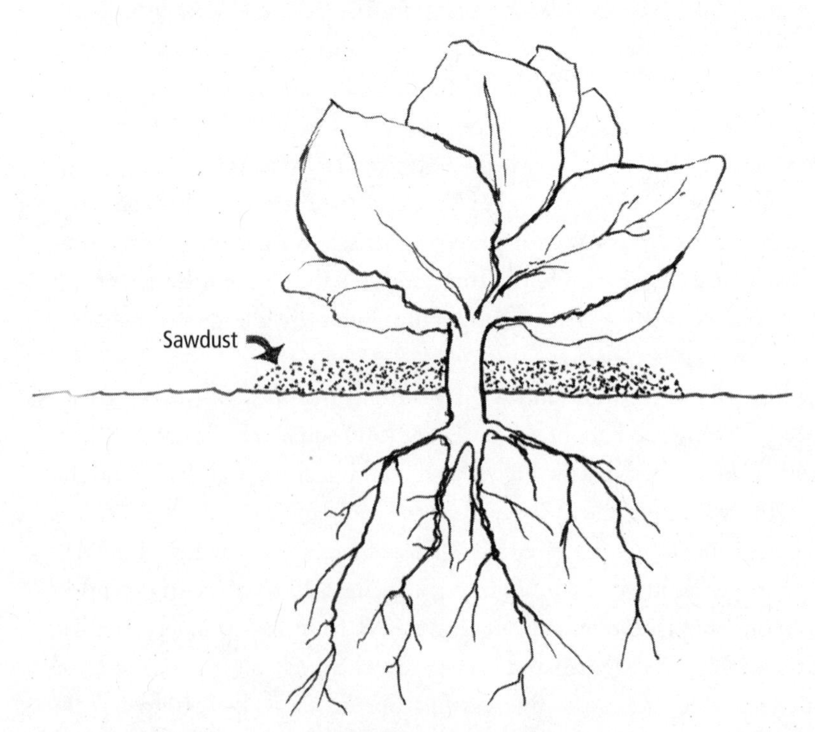

Figure 8.2: *A sawdust collar that protects brassicas from root maggots.*

Floating row covers that are carefully anchored on all sides will effectively prevent cabbage root maggots from reaching plants. These are especially useful for Chinese cabbage and turnips.

The carrot fly maggot is a similar Cascadian (and British) pest that gardeners in eastern North America are fortunate to rarely encounter. Although I gardened in Oregon for many years, I have little personal experience with this critter, although I know people who lived only 40 miles (65 kilometers) from my garden who had serious problems. The fly begins breeding in late summer, going through a generation every month and becoming a plague by winter if the season is not too severe. Carrots started in late May come up after the spring hatch is through; these may finish growing relatively unharmed and be harvested by late summer. However, carrots left in the ground after summer ends become increasingly infested as the fly population grows a hundredfold or more each month into autumn and early winter.

Parasitic nematodes are useless because this maggot is most active when soil temperatures are low and nematodes are inactive. Covering the bed with meticulously anchored spun-fiber row covers may give you a maggot-free crop. When the tops are three or four inches (7.5 to 10 centimeters) tall, thoroughly thin the carrots to about 150 percent of their normal spacing (to allow for the loss of light through the fabric) and at the same time eliminate every weed, then carefully cover the bed. Gardeners have used homemade solutions similar to this for years, placing large, lightweight, wooden frame boxes covered with fly screen over carrot crops to prevent the fly from laying eggs.

Store carrots during winter by carefully laying sheets of plastic over the carrot tops and covering that with a few inches of straw or soil for insulation. This might prevent the fly from gaining access, through it may also make a haven for field mice, who enjoy carrots as much as humans do. This works in Cascadia because severe freezing conditions are rare, and when they do happen, they normally last only a few days.

Slugs and snails *(Gastropodae)*

See *Growing Vegetables West of the Cascades.*

Symphylans *(Scutigerella immaculata)*

See *Growing Vegetables West of the Cascades.*

Squash borer *(Melittia satyriniformis)*

Found everywhere in North America east of the Rockies, the borer can be highly destructive because a single one can cause the loss of an entire plant or collapse a long runner. The adult moth emerges from the soil in spring and lays eggs singly near the bases of numerous squash plants, continuing until she has laid as many as 250 eggs. A few dozen moths are capable of infecting an entire commercial pumpkin field. Hatched larvae enter the plant, usually near the base (the entrance hole is marked by a gob of what resembles sawdust), and inevitably tunnel "upstream" toward the center, destroying the vascular system and causing that runner (or the whole plant) to wilt and die. Occasionally one enters a leaf stem. Tunnelling goes on for four to six weeks. If the vine dies before the borer completes its life cycle, it can migrate to another plant and start again. Because it is protected inside the vine, a borer is difficult to eliminate with organic pesticides. I suspect the borer is the reason that squash is traditionally grown several plants per hill — as insurance — even though all cucurbits grow much better if only one plant occupies each hill.

The borer prefers squash and pumpkins; usually it has little interest in cucumbers or melons. Butternuts, Green-Striped cushaw, Yellow Crookneck, and Dickinson pumpkin all have proven resistance to the borer. Gardeners should patrol their plants every few days and carefully inspect them for signs of borers, although it can be difficult to closely examine bush summer squash of any size. For that reason the sprawling heirloom Yellow Crookneck might be the best for the garden; that variety also shows the highest resistance to the cucumber beetle.

If you discover signs that a borer has entered a plant, you can destroy it by slitting the vine, starting at the entry hole and cutting toward the center of

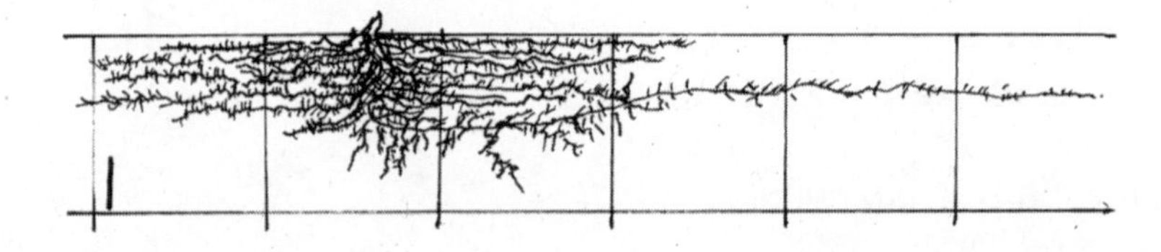

Figure 8.3: *Typical secondary root originating from the leaf node of a squash vine. It is only partially developed. Each square is one foot (30 cm) per side. (Graphic adapted from Figure 89 of John E. Weaver's* Root Development of Vegetable Crops.*)*

the plant. If it is a vining squash, toss a shovelful of soil on top of the vine at the cut to accelerate healing and cause secondary roots to form there. Some gardeners routinely toss a shovelful of soil atop many points where leaves emerge from vines. This encourages secondary roots to form at each point, so if a borer does manage to invade, it cannot collapse the entire runner or entire plant because it will have multiple root systems. Do not miss any opportunity to destroy a borer; there usually aren't many adult moths causing all the trouble.

Plot rotation does not help because borer moths may fly as far as half a mile (800 meters) to lay eggs. After harvest, promptly burn winter squash vines or move them to the core of a hot compost heap to prevent late-maturing borers from pupating. If your growing season is long enough, later planting may allow you to evade the moths, which only fly and lay eggs for a short period. The local agricultural extension office may suggest a safer sowing date. Having fertile soil under the entire spread of the vine is helpful, as a vigorously growing vine can tolerate a few borers, especially if it forms secondary roots in many places. Some gardeners use a large syringe to inject *Bacillus thuringiensis* var. kurstaki into the stem of the plant. In an attempt to kill borers emerging from the soil before they get into the plant, some gardeners repeatedly spray rotenone/pyrethrum, confining it to the base of the stem. Do this late in the day because this spray combination is highly toxic to pollinating bees.

Squash bug *(Anasa tristis)*

Squash bugs are easily recognized. They're about five eighths of an inch (1.5 centimeters) long, dark brown or mottled, and give off an unpleasant odor. Children call them stinkbugs. The adults and their larvae suck plant juice and inject a toxin that causes plants to wilt. Individual runners may blacken and die back. The effect resembles a bacterial wilt, but when the bugs are brought under control, the plants recover. They are considered the most destructive North American cucurbit pest.

Adults overwinter under whatever cover is available, emerge in spring, and begin feeding and laying eggs. The eggs are easy to spot: orange-yellow, a sixteenth of an inch (1.5 millimeters) long, placed in neat rows on the undersides of leaves. Upon hatching, the larvae start feeding. Egg-laying can continue well into the summer. In the south, squash bugs can go through two generations each summer, increasing to huge populations.

The first step to reduce the problem is end-of-season sanitation, leaving as little cover (such as old boards on the ground and dead plant residue) as possible. Prompt and thorough composting of all squash vegetation in the core of a hot compost heap is helpful. Also dig up the root mass at the center of the vine and compost that, too. The home gardener's remedy is to thoroughly and repeatedly handpick stinkbugs. Remove and burn leaves showing eggs, or zip them up in sealed garbage bags.

The amount of time these actions take would make most people think of pesticides, but in this case pesticides are not useful. Biological pesticides are effective only against small-sized larvae, and because egg-laying goes on for several months, because most organic pesticides last only a short time, and because the vines grow rapidly, these pesticides have to be repeatedly sprayed *on the undersides of leaves* (in itself not an easy task to accomplish). Chemical pesticides kill bees pollinating the flowers. There are numerous beneficials that eat squash bugs or their larvae; spraying poisons kills them off, too.

The gardener's best options are to grow a mixed garden offering lots of cover for beneficials, including some stands of buckwheat to harbor tachinid fly parasites, and handpicking.

Squash bugs have varietal preferences, though research reports on these are mixed and somewhat contradictory. Gardeners will have to work this out for themselves.

Diseases and their remedies

Sadly, the home gardener has no effective cure for plant diseases. However, we can often prevent the problem by growing healthy and naturally disease-resistant plants, which means using appropriate varieties and making highly fertile soil. If disease does strike, keep in mind that some years the weather can be so unfavorable for certain types of vegetables that almost any variety will be weakened to the point of becoming sick.

The best first step when some plants are getting diseased or when the weather turns against you is to spray your garden with liquid kelp, which provides a tonic of trace minerals and other fortifying substances that improve plants' overall health. In the same vein, Dr. Elaine Ingham, an Australian at Oregon State University, has been spraying teas brewed from high-quality composts. She finds compost teas cover the plant with beneficial microorganisms, and for a week or so after spraying they will prevent disease organisms from gaining a toehold. However, this practice is not simply a matter of dumping any old compost into a barrel and brewing up some "tea," filtering it so it won't clog the sprayer's nozzle. The compost has to be quality stuff, expertly made. The ratio of manure to vegetation, and the quality of that vegetation (i.e., how much woody matter does it contain and how much green stuff), may determine which diseases the tea will control. Still, if you make decent compost, it is worth a try to see if yours will serve this purpose (see Chapter 6 for a compost tea recipe). If you make good compost, you might want to do an Internet search for "Elaine Ingham" or "compost tea" and see what comes up.

Regular (weekly) foliar feeding of vegetables is a good idea in any case. It doesn't cost much. Every gardener should experiment using homebrewed compost/manure teas, liquid kelp, and fish emulsion sprays. Those who don't practice the organic faith might find great benefit from spraying hydroponic nutrient solutions as foliars.

Powdery Mildew

Powdery mildew (PMD) covers the photosynthetic surfaces of the leaf with a grey cast and interferes with food making. It quickly kills most plants it attacks.

Here is one case where the gardener has an effective way to stop the disease in its tracks; however, the remedy is quite short-lasting and requires painstaking attention. Spray infected plants with a solution of one measured teaspoonful (5 milliliters) of baking soda per quart (liter) of water, adding just enough liquid soap or dishwashing detergent to insure the droplets spread out and cover the leaf instead of beading up and running off. A squirt of this will instantly kill the mildew. However, the disease usually occurs when conditions are unfavorable to the plant, and after you've sprayed it away, PMD will usually appear again in a few days. If PMD occurs during a short spell of unseasonably unfavorable weather, stopping it for a few days can save the day.

I suggest mixing kelp tea and fish emulsion into the spray tank with the baking soda/soap solution. Why not.

CHAPTER 9

What to grow ... and how to grow it

This chapter provides growing details, vegetable by vegetable. It is not organized A to Z. Instead, the vegetables are listed in an order based on their importance to a self-sufficient homestead economy and their difficulty to grow. That's why kale and potatoes (both Irish and sweet) come at the beginning and celery comes at the end. I am only going to provide the essentials and leave it to experience to reveal the rest to you. I will assume you have understood what I said in previous chapters. If you find yourself puzzled by anything that follows, I suggest you do a bit of rereading.

Some general tips

Sowing depth. Tiny seeds like celery, basil, sorrel, and most of the herbs should fall into tiny cracks and crevasses of newly raked soil, to be barely covered (if at all) with a sprinkling of fine compost. Then press the earth down gently either with your hand or the back of a spade, much as you would roll a large area like a lawn. This essential step restores capillarity. Fine seeds can only be directly seeded outdoors in mild temperatures, or the rows must be shaded temporarily until they sprout. Ordinary small seeds the size of brassica, carrot, parsley, and fennel are sown about half an inch (1.25 centimeters) deep. Larger "small" seeds like spinach, beet, chard, radish, and okra are sown about three quarters of an inch (two centimeters) deep. Large seeds like the legumes, corn, and cucurbits are usually planted to a depth of about four times their largest dimension.

Vegetables discussed in Chapter 9

Fertility needs. I assume that you understood the three levels of soil fertility I set out in Chapter 2, where I described the idea of low-, medium-, and high-demand vegetables and showed you how to create minimum levels of soil fertility to grow them. Don't forget that these are minimums; all vegetables grow much better when given soil more fertile than the bare minimum.

Plant spacing. I will not be repeating the information on spacing provided in Figure 6.1. I suggest you flag that page or make a photocopy of it for rapid access. In this chapter, I'll refer to various spacing arrangements in shorthand. The term "on stations" is British. It means that a few seeds are sown in a cluster, usually on raised beds or wide raised rows. The clusters are at fixed distances, such as 24 by 24 inches (60 by 60 centimeters), and each cluster is thinned progressively. Sowing seeds "in drills" means they are set in the bottom of a furrow. Sometimes seeds are to be sown in highly fertile hills. These hills can be made in a raised bed on stations, sometimes not.

Progressive thinning. I will frequently suggest that you thin to a single plant, progressively. Here's how it works. Suppose you're growing looseleaf lettuce and the mature heads will need to be 12 inches (30 centimeters) apart.

First sprinkle seeds thinly in drills. Ideally, if you get excellent germination, you will have about one seedling emerging every inch (2.5 centimeters). However, distribution this uniform doesn't happen unless you're a farmer using precision planting equipment and sowing the highest-quality seed of predetermined sprouting ability. In the home garden, clumps of seedlings inevitably emerge in places, so immediately after emergence you should reduce severe competition: thin any seedling clusters so the survivors don't quite touch. When this initial plant density starts to compete for light (the seedlings will lean away from each other) thin them again so they don't quite touch. A week or ten days later, when they are again bumping, take out every other plant. Suppose at this point they are three inches (eight centimeters) apart. When these are touching, remove every other plant. Now they're six inches (16 centimeters) apart. When these plants are touching, cut off every other one again. Thinnings of this size are definitely salad material. And now the survivors in that row are properly spaced to reach maturity.

Seedling clusters on stations can also be thinned progressively. Start with three to five seedlings and gradually reduce their number so that by the time the best plant has a few true leaves and is securely established, it stands alone on that station.

Root system drawings. Muriel Chen, my wife, copied the drawings of root systems in this chapter from Weaver's classic study *Root Development of Vegetable Crops* (RDVC). They are included to help you realize that the plant growth you can't see is as important as what you can see. Each drawing's scale is one foot per square. Keep in mind that Weaver worked in a rather dry climate on deep open soil that posed little opposition to full root development. In more humid climates, most soils divide into layers called topsoil and subsoil. The subsoil is typically clayey and won't allow such full expression of root development — but you can be assured that the roots are trying to develop further, nonetheless.

Weaver's drawings, combined with an understanding of how roots work, will help you garden better. Almost every plant tries to control the soil below it. To do that it secretes chemicals that repel the roots of other species. Some of these chemicals are so effective (and long-lasting) that a year after one crop is grown in a spot, another species may do poorly because these root exudates are still present.

Also, notice in Weaver's pictures that roots never turn back and grow toward the center. The plant tries to penetrate new soil for untapped sources of moisture and nutrition; it would be a waste to make a denser root system than necessary. How do plants "know"? Again, it is root exudates — chemical signals stop its roots from growing near its own already-established roots.

Finally, and this fact may be the single most important thing you'll ever grasp about growing plants, **for a plant to acquire nutrition efficiently, it must have an ever-expanding root system.** A root extends only from its tip, and it is capable of efficiently assimilating moisture and nutrients for only a fraction of an inch behind the tip. That's because a few days after it is formed, what was the root tip becomes covered by a kind of bark that reduces penetration of moisture. It is only by creating new root tips in an ever-increasing and ever-expanding network that the plant can feed efficiently. But the plant can't readily make new root tips in areas it has already filled with roots. And because of exudate warfare, it can't make them effectively in areas that another plant has already filled with roots. My point is that **when root systems begin to compete, the plants are not as able to acquire nutrients.** This is a stress to them, and they may show it in various ways: their growth will slow, they will be more easily attacked by insects; they will stop producing as much new fruit; they will be more susceptible to diseases. They can be starving in the midst of plenty.

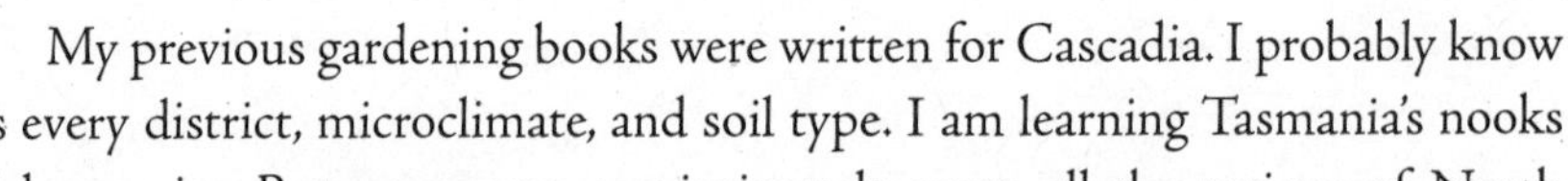

My previous gardening books were written for Cascadia. I probably know its every district, microclimate, and soil type. I am learning Tasmania's nooks and crannies. But no person can intimately grasp all the regions of North America, much less the rest of the English-speaking world. So in this section I usually won't be able to tell you three things: (1) precise planting dates for your location; (2) the most suitable varieties for your district; and (3) the handling of some pest or disease problem either unknown to me or unique to your area. These kinds of data are best obtained from a governmental agricultural advising service. You're paying the taxes to support it; make use of it, and give those civil servants a reason to draw their salaries. I hope that the third point will be an infrequent event if you take my advice about making plants healthy by making soil fertile and by not asking a plant to handle seasons or soil conditions it is not bred for.

Crops that are easiest to grow

In this section I cover vegetables that are generally trouble-free and low- to medium-demand in terms of soil fertility needs. To keep the size (and cost) of this book lower, I discuss a few harder-to-grow relatives along with an easier one. New gardeners should not bet the ranch on more-demanding veggies than the ones in this section.

Kale, collards, and giant kohlrabi

There are two kale species. One is a *Brassica oleracea.* I'm not going to be routinely tossing Latin names around in this chapter, but in the case of the brassica family the distinction is useful because the other kale species, *Brassica napa* (it includes rutabagas), grows in an entirely different manner. The *B. napa* kale is usually called Siberian. *B. oleracea* grows an unbranched tall central stalk. Siberian forms a rosette pattern, meaning that all the leaves come out of a central point close to the ground, similar to a lettuce or spinach plant. Some prefer Siberian's flavor when kale is used raw in salads.

Kale of either sort is the most vigorous and most cold-hardy of all garden brassicas. It will produce when other coles fail. Collards are a non-heading cabbage only slightly less vigorous than kale and may be grown just like kale. Giant kohlrabi is basically a low-demand fodder crop whose tasty globe can grow to the size of a volleyball.

All coles need more calcium than most other vegetable species. If you're not using COF and if you are gardening on rain-leached land (where there is enough rainfall to grow a lush native forest), then before sowing any brassica crop you should broadcast and work in five pounds (about a quart or 2.25 kilograms/ one liter) of finely ground agricultural lime per 100 square feet (ten square meters) of growing area. Do not spread more lime than that without a soil test.

I have seen kale resume growing after thawing out from an overnight low of 6°F (–14°C). As long as the soil has not frozen, kale will keep going. So will endive, spinach, and a few other minor salad veggies. Having fresh salad greens throughout the winter will certainly appeal to any gardener living where the snow lies thickly. I refer those living where the soil freezes in winter to Eliot Coleman's book *The Four-Season Harvest,* listed in the Bibliography.

Growing details. Kale's flavor gets sweeter after some frost in the same way the chicories do. Collards don't need chilling to be good eating, which is

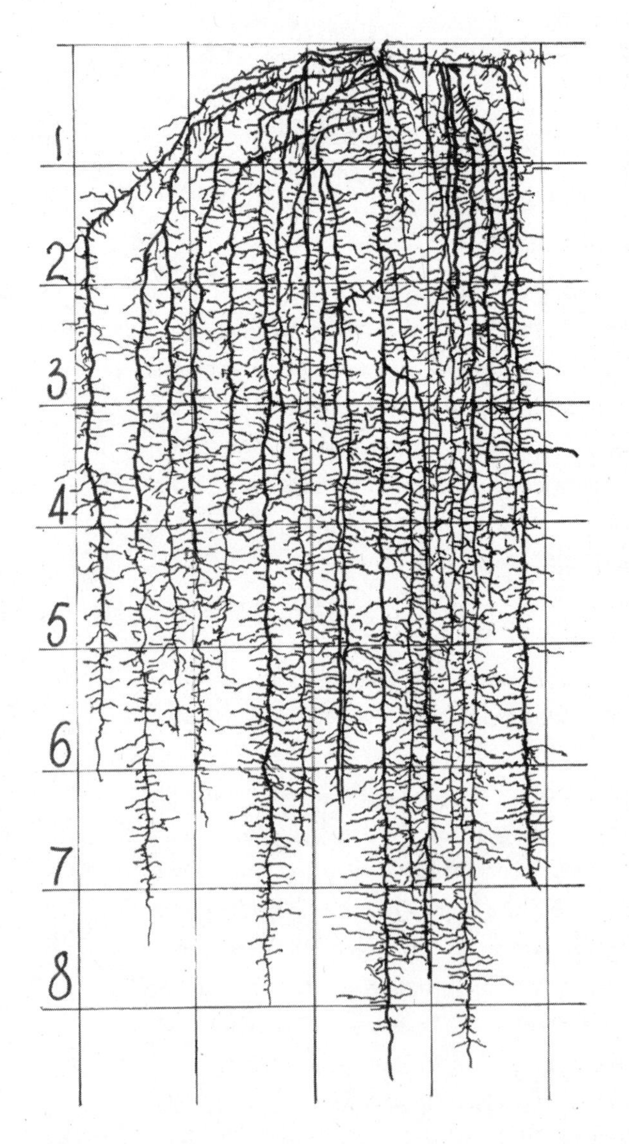

Figure 9.1: *This kohlrabi was sown in spring; the drawing is made after 120 days of growth. Clearly this is a vegetable capable of foraging. Kale's root development is not very different. (Figure 40a in RDVC)*

probably why they are more popular in the American South. I enjoy seeing huge frilly plants by autumn, so I start kale about three months before the first frost on 24- by 30-inch (60- by 75-centimeter) stations. I once grew giant kale by direct-seeding it in spring on four-foot (120-centimeter) centers and initially making their soil as fertile as possible. Each one reached four feet in diameter and 4½ feet (140 centimeters) high by frost. Had I given them any water at all that summer they probably would have become as tall as I am.

As a hasty fill-in crop started six weeks before the first frosts, kale will make a small plant and in this circumstance should be sown in drills about 12 inches (30 centimeters) apart and thinned progressively. Whether it's huge or modest, it'll be equally frost-hardy. No extra fertility should be needed for late sowings as the kale will make use of what is left from the previous crop.

I start giant kohlrabi just after the solstice for harvest as an autumn and winter crop. It is grown as if it were a medium-sized cabbage. Although it will produce at fertility levels suitable to a field crop, giant kohlrabi will definitely respond to better soil and abundant soil moisture by becoming larger, more mildly flavored, and more tender.

Pests and diseases. I always am pleased to see that cabbage worms do not have any interest in my kale. If they did I would spray Bt. Kohlrabi is only slightly more interesting to them. Perhaps this disinterest is because the moths are diverted to the more refined brassicas in my garden — which serve as a trap crop and certainly do require regular spraying with Bt.

Growing refined brassicas. If you can grow a large kale plant, then you have also mastered the art of growing the weaker, more inbred, large brassicas; all were bred from the same wild *Brassica oleracea* that kale came from. Imagine a head of cabbage as a non-curly kale whose stalk has become so shortened that it has next to no space between leaves and that overemphasizes kale's tendency to wrap its leaves at the growing point. The Brussels sprout is but a kale that makes little cabbages at each leaf joint. For broccoli and cauliflower, imagine a slightly less aggressive (and less cold-hardy) wild oleracea strain that naturally made larger flowers; now it is bred to make enormous flowers. As you move up the scale of refinement from kale to cabbage to Brussels sprouts to broccoli to cauliflower, each requires more fertile and more open soil and even more moisture than the previous one.

One last comment on refinement: the cabbage is intrinsically a large vigorous plant that makes a hefty head. However, to create little heads currently popular with the supermarket trade, cabbage has to be highly inbred and weakened. These small-framed varieties are touchy little critters with delicate root systems intolerant of dry or compacted soils. They need to be coddled. New gardeners will do better to grow bigger ones.

All brassicas may be directly seeded exactly as I suggest for growing kale. It is best to make small hills in raised beds for the refined brassicas, concentrating an extra half cup (120 milliliters) of COF or an extra large double handful of strong compost immediately under their stations. Two points about cauliflower: it has a particularly weak root system that doesn't thrive in clayey soils. It also does not like maturing in heat, making spring cauli sowings a bit chancy because of the unpredictable onset of hot weather. A few broccoli varieties may handle hot weather better; the catalog will proudly state this. The safest thing is to schedule cauliflower for autumn harvest. In the mild-winter climates, refined brassicas may be started after the heat of summer

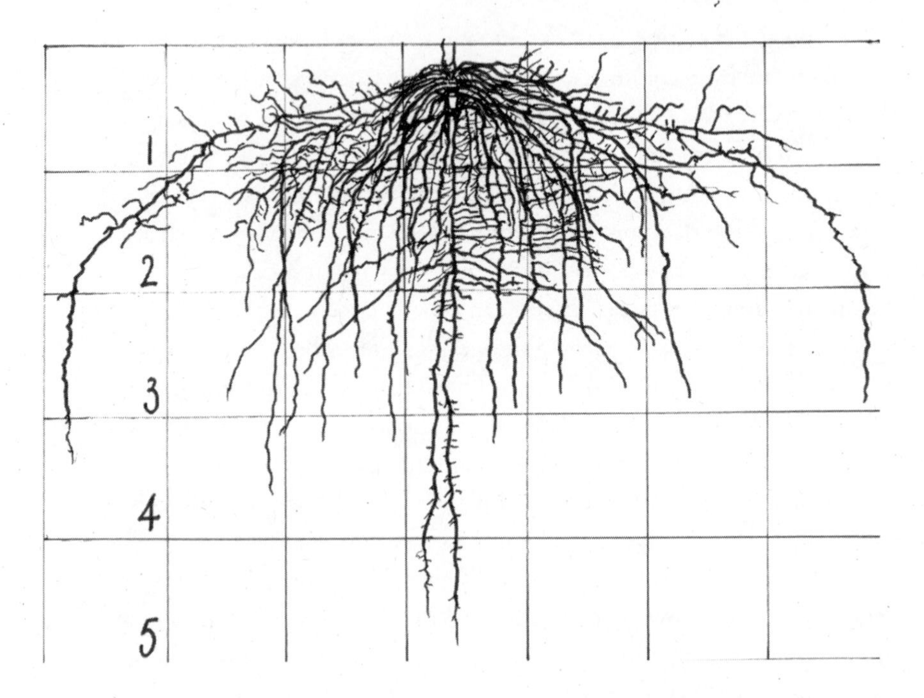

Figure 9.2: *The root system of a large-framed midseason cabbage after growing about 100 days from direct seeding. The leaves have not yet wrapped into a head. Harvest will be in another four to five weeks. At harvest the root system of this plant will thickly fill the soil to a depth of five feet (150 centimeters) but will not become more extensive. Kale makes a more extensive and deeper root system than this. (Figure 29 in RDVC)*

lessens for autumn and winter harvest and, by using the right varieties, for overwintering and spring harvest.

Varieties. Winterbor (hybrid) is the most frequently offered oleracea type because it is more vigorous and perfectly uniform. Open-pollinated (OP) kale is vigorous too, so the old OPs will serve fine. At the time of writing, though, when it comes to refined brassicas there are no productive OP Brussels sprouts left except in Chase's catalog (see Chapter 5).

Harvest, storage, and use. With kale, if you do not pluck the growing point at the top of the stalk, the production of new leaves will continue. If olracea-type plants overwinter, in early spring they will begin making numerous small (more tender and more delicious) leaves all along the thick woody stalk. (You'll think of Brussels sprout when you see them.) Collards are often sown in drills with the intention of eating the progressive thinnings. Kale can be used this way, too. Kohlrabi will await harvest through rather severe frosts; where the ground freezes it will store well in the root cellar.

Kale and collards are mainly used as pot greens. To encourage those unfamiliar with eating them, may I recommend a simple and (naturally) frugal Scottish recipe called colcannon. Fill a large pot with finely chopped kale, add a quarter inch (six millimeters) of water, and set the heat at low. As soon as the leaves collapse, add an inch-thick (2.5-centimeter) layer of roughly cut up, unpeeled potatoes and allow the whole thing to steam until the potatoes are soft enough to mash. Then mash everything. Do not pour off any remaining water. Mash the lot and retain the minerals. Add your usual mashed-potato seasonings. Steamed kale by itself is a bit intense. And potatoes by themselves are usually a bit low in protein and minerals. Combined they make near-perfect nutrition with complementary flavors.

Finely shredded in moderate quantities, kale (especially Siberian) blends well into greens salads. Kohlrabi is tasty when coarsely grated and made into slaw-type salads as though it were cabbage. I like dipping raw kohlrabi chunks.

Saving seed. All *Brassica olracea* cross-pollinate; bees do this task. Crosses are unlikely to make desirable plants. Isolating different sorts by a half mile (800 meters) may do for low-quality seed. Siberian kale, also bee-pollinated, crosses only with rutabaga (swede) and has the same relationship to the rutabaga as Swiss chard (silverbeet) has to beet (beetroot) — it is a rutabaga bred for tasty leaves instead of a bulbous, flavorsome root. Otherwise, Siberian makes seed like any other brassica.

These brassicas are biennial, meaning they must pass through a season of cold weather and short daylength before flowering is triggered in spring. To overwinter large kale plants with thick woody stalks where the soil freezes, dig them up carefully in late autumn so as to preserve much of their root system, replant them in a bed of damp soil in a root cellar, and let them rest there until spring, when they are transplanted back outside. They might survive a not-too-severe winter if buried under enough soil and then exposed again in spring; cabbage seed is made this way in Denmark. In milder locations they survive winter unprotected.

The blooming plant makes huge floral sprays. Each of the thousands of small yellow flowers makes a thin pointed pod that holds a few round black seeds. The earliest time you should harvest is when some of the earliest ripening pods have shattered (released seed) and most of the rest contain ripe or nearly ripe seed (the seed is fully ripe when it has turned dark brown or black). Pull the plants, roots and all, shake off as much soil as possible from the roots, and spread them on top of a (big) tarp in the shade under cover, where there is good airflow, to dry slowly and finish ripening. Then on a bright warm day of late summer, drag the tarp into the sun and let the whole lot of straw dry to a crisp. Late that afternoon do some marching in place atop the straw, releasing most of the seed from most of the pods. Lift off the strawy bits and put them in the heap of dry vegetation awaiting your next compost pile. Left on the tarp will be seed, broken seed pods, and some smaller trash. Put it all in a large bucket and then, in a light breeze, slowly pour the contents of one bucket into another. Hold the top bucket a few feet above the bottom one so the breeze may blow away the light stuff while the seed falls into the lower bucket. This process is called winnowing. You may have to pour from bucket to bucket numerous times before you have relatively pure seed free of chaff. (It can help to screen out the larger bits of chaff before winnowing; use any sort of improvised sieve.) Do not be concerned if you lose up to a third of the seed due to wind blowing it beyond the bucket waiting to receive it. This is the lightweight unripe stuff that will have poor storage life and low germination — the seedroom floor sweepings.

Do not attempt to save either sort of kale seed unless you have at least six plants involved in sharing pollen. If you work from too small a plant population, you'll experience a rapid onset of inbreeding vigor depression. Kale or collards are the only large brassicas the home gardener should ever attempt to grow seed for. I would not advise growing seed for the refined large brassicas unless you have at least 50 plants in the gene pool — 200 is better. But with kale, involving such a small number of plants is okay because kale still retains most of the vigor of wild cabbage; even if some of that vigor is lost, it'll still grow okay.

Potatoes (Irish)

Introduced to Europe after the Incan conquest, the potato languished for around two centuries. Because these initial varieties were adapted to tropical daylengths, the potato was considered a low-yielding curiosity. Eventually, better varieties were bred. Then the potato caused a European social revolution because it produces many times more actual nutrition per acre than any other staple crop except perhaps paddy rice. The potato allowed a cottager with less than an acre to feed a family. So Europe's population increased rapidly. This is one reason there were so many European peasants coming to the United States after the War Between the States.

Potatoes need not be merely starch. They can contain up to about 11 percent protein (dry weight), matching the protein content of human breast milk. The quality of nutrition you end up with has a lot to do with both variety and the pattern of soil fertility. If you're growing a starchy variety, which is flaky and crumbly (and often called a "chipper" because it is the starch that browns nicely when making potato chips), and if you are growing your spuds with lots of moisture and fertilizing your soil so that it offers the plant excesses of potassium, then you'll end up with a much bulkier harvest of low-protein spuds. If you grow a "boiling variety," which is often yellow-fleshed with a waxy structure that doesn't fall apart when boiled, if you reduce or avoid irrigating once tuber formation begins, and if you build your soil's fertility so that it has considerable mineral nutrients but a rather low level of potassium, you'll end up with a somewhat smaller bulk yield of somewhat smaller-sized spuds that have considerably more taste and nutrition.

Growing details. The "Irish" potato is grown much the same way in all variations of temperate climates. The vines are not frost-hardy, so it must be planted late enough in spring to emerge after the last frost. Frost is not a complete risk; if the young vines are burned by frost, more will emerge. But the loss of the first bit of leaf will reduce the ultimate yield, so it's best to avoid it. In mild-winter (frost-free) regions

To achieve nearly complete food self-sufficiency using European cereal grains, a family needs more than an acre and will almost have to use draft animals (which will need food grown for them, too, requiring working even more land than the first acre) or a husky rotary cultivator. If the staples are the Native American ones — corn, beans, squash, and sunflowers — an acre garden that can be worked entirely by hand labor will serve an extended family. But if the staff of life is the lowly spud, far less land than that will serve, considering that potato yields generally exceed 300 bushels per acre (10,000 kilograms per hectare). ■

it is possible to sow late in summer and grow spuds as a cool-season crop, but this practice results in low productivity. That's because the tuber is a savings account of surplus sugar made by the leaves, and when intensity of sunlight drops off markedly after midsummer, so too does sugar formation decline. Despite the low yields, winter spuds are grown in frostless Florida because of the high prices obtained for new potatoes in the north during late winter and early spring.

Root cellaring

To maintain a body in robust health you must feed it a sizeable amount of fresh food, preferably raw. The nutritional quality of canned and frozen foods has been massively reduced. The same is almost as true of dried foods, especially if they were blanched during processing. Fortunately, in cold-winter regions it is possible to store fresh vegetables and fruit in living condition for many months without using any energy to do so. This is accomplished by cellaring. Imagine having the makings for a fresh leafy green salad in the cellar throughout the winter; eating bins of root vegetables, your own cabbages, and perhaps Brussels sprouts (still on the stalk) in midwinter; or sprouting your own Belgian endive and not considering it an expensive delicacy.

Few North American homes are still equipped for root cellaring because most of them have the furnace in the basement. However, it may be possible to wall off and highly insulate a part of the basement for use as a winter food-storage cellar. Otherwise you can dig a cellar outside, though this may be less accessible during the time of heavy snows. Making a root cellar is not generally regulated by building codes and requires no permits nor adherence to any prescribed construction techniques (other than the requirement that it not collapse while you are inside it). Cellars may be made of recycled materials. Even old chest freezers may be recycled into small root-storage compartments.

The basic technique behind cellaring is to rapidly lower the temperature to a few degrees above freezing and then hold it there, steadily, through the winter. During autumn you open air vents at night and close them during the day. During winter, less ventilation is needed because, above all, the cellar must not go below freezing. The more stable the temperature, the better the food in the cellar will keep.

Using the term "root cellar" shows the limited application most Americans made of this procedure. The Europeans have demonstrated far more ingenuity and wouldn't dream of putting only root ☞

"Seed" consists of chunks cut from large potatoes or whole small ones. Vines emerge from the eyes. The ideal seed is termed a "single drop," a small uncut potato weighing about two ounces (60 grams) and having at least two eyes. Each chunk cut from larger potatoes must also have at least two eyes and weigh at least two ounces.

Farmers, using machinery, must plant unsprouted seed pieces because any sprouts would be knocked off by rough handling. So farmers usually treat the

crops and apples in storage for wintertime. Basically, roots are put in slatted boxes on shelves or, if the vegetable has a tendency to dry out (like carrots or beets), are packed in damp coarse sand and housed in barrels (or large trash bins). Leafy crops like cabbages and heads of endive and escarole, which blanch during storage, becoming milder and more tender, may also be kept over winter. Immediately before it gets too cold for them outside, which is after some frosts but before hard freezing begins, dig them up, shake the earth from their roots (sometimes you need to break off the large outer leaves), and then transplant them into the cellar and put their roots into shallow beds of soft moist soil. There is no reason why such earth beds could not be constructed on a concrete slab, although traditionally the floors of such cellars were bare earth.

The only way people in snow country can make seed for most biennial crops is to cellar them over winter and then plant them back outside in spring. This was done extensively on a home-garden scale (and commercially to a lesser extent) in the United States before the West Coast seed industry developed.

Root Cellaring, by Mike and Nancy Bubel, is an excellent book about cellaring that is worthy of the most profound consideration.

I mention cellaring for those of my readers who may need to make use of it. However, since being an adult I have lived in climates where the technique is not applicable. In California, where I first learned to garden, veggies grew lushly 12 months a year. In Oregon, a well-insulated, unheated outbuilding was plenty good enough for spuds and apples, while carrots and other root crops would overwinter in their growing beds with only a bit of soil put atop them to protect their crowns from a chance shallow freeze. And greens! Green salads grew right through the frosts and even emerged from the occasional quick-melting snows to grow some more. Ah Cascadia! Closest thing to paradise there is in North America. ■

cut surfaces of their seed-potato chunks with fungicides because they'll be in the earth and subject to rotting for some time before the eyes start growing. Home gardeners can do much better. We can chit our seed and then handle already sprouting seed gently enough that the shoots aren't damaged. Chitted seed is already growing when it is planted.

About six weeks before planting, spread uncut seed potatoes on a tray in a brightly lit room that is not too well heated. I put mine on cookie sheets in front of a window that gets no direct sunlight in order to avoid drying out my seed. By planting time the potatoes will have turned light green, and shoots will be emerging from many of the eyes. On planting day I cut the larger potatoes in chunks; I try to make sure each chunk contain two eyes that are actually sprouting.

A day or two before planting, dig the rows. One row of potatoes can make luxurious and extraordinarily high-yielding use of an entire four-foot-wide (120-centimeter) raised bed when planted longways down the center, but if you're growing more than one row, it's best to plant spuds on flat ground, making their long rows about 36 inches apart on center (90 centimeters). Spread compost or well-rotted manure atop multiple rows in one-foot-wide (30-centimeter) bands, a quarter to a half inch (6 to 12 millimeters) thick, and then deeply dig the compost-covered rows (not the spaces between the rows). Try to loosen the earth well. If you want the highest possible production and aren't growing a huge plot, excavate the rows first, before improving their fertility. Remove soil one shovel blade wide nearly to the depth of a shovel, set it beside the ditch you're creating, put fertilizer and/or strong compost into the ditch, and then, standing in the ditch, dig it in as deeply as another full shovel's length. That'll put the fertility well below the seed piece, where the main root system will form.

I do it this way: I spread four to six quarts (four to six liters) of a lime-free complete organic fertilizer per 50 row feet (15 meters) into that shallow ditch and dig it in well. Then I spread a dusting of well-rotted manure or compost over the soil that was removed from the trench and push it all back into the ditch. I end up with a low mound above a zone of loose, humusy soil about a foot (30 centimeters) wide and a foot deep. Below that is another zone of highly fertile soil. If I am really shooting for the highest possible yield, and if I have the free time and energy, after the seed has been planted but before it starts putting roots out I will spade up the earth between the rows.

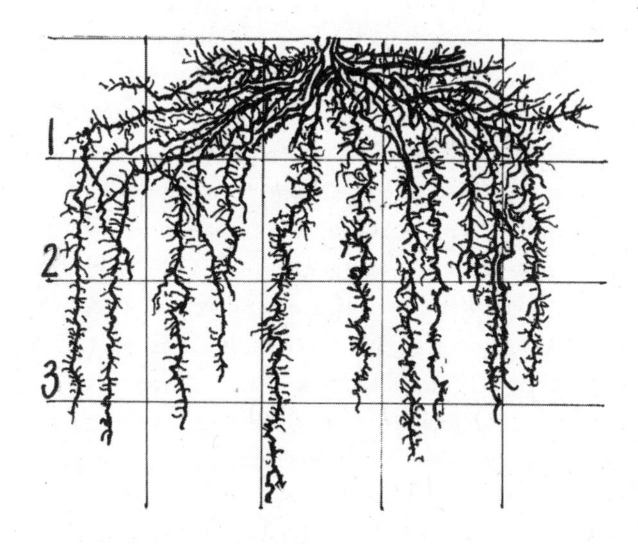

Figure 9.3: *A potato plant that enjoyed good soil moisture in deep open loam, shown at the point of beginning to form tubers. The spacing that produced these roots was 14 inches (35 centimeters) apart in rows three feet (90 centimeters) apart, which ideally allows each plant to be nearly the sole occupant of its own root zone. (Figure 40 in RDVC)*

To plant sprouting potatoes, pull back the soil from a spot in the center of the row with your hand, opening a small hole about four inches (10 centimeters) deep. Gently set the seed in. If there are long shoots, point them up. Then cover it up. Space the seed pieces 8 to 12 inches (20 to 30 centimeters) apart. Wider spacing will give you more lunkers (huge specimens), a slightly lower yield, and a better ability to handle dry soil. Closer spacing will give you moderately sized potatoes, the largest possible yield, and less ability to handle dry spells.

Farmers must plant their seeds deeply in loose soil and let them grow, accepting that a portion of the spuds forming at the surface will green up and be inedible. When potatoes are grown this way, the soil tends to settle and become compact, reducing tuber formation. Gardeners can do better; we can plant our seed shallowly and then hill up our plants as they grow, thus providing a looser medium for the potatoes to form in. We end up with a higher yield of smoother spuds, while at the same time thoroughly eliminating weeds. Keep in mind that all potatoes will form above the seed piece and that if you do as I am suggesting, the seed will have been placed only a few inches below the original soil line, so your crop will be easier to dig.

A few weeks after planting, the vines appear, emerging more quickly than they would if you had put the seed deeper because the soil near the surface is warmer in spring. After the vines have grown about four inches (10 centimeters), start hilling them up. Using an ordinary hoe, walk beside one row and, reaching over it with the hoe, scrape up a bit of soil from between it and the next row and pull that loose soil (and any weeds you cut off with your super-sharp hoe) up against the vines. Bury the bottom inch (2.5 centimeters) of the vine. Do that from both sides of the row. Five to seven days later the vines will have grown another few inches. Hill them up another inch or two. *Never cover more than a quarter of the new growth.* By the time the vines are blooming, you should have formed an 18-inch-wide (45-centimeter) mound of loose earth about ten inches (25 centimeters) tall with the vines emerging from the center. It is in this mound that almost all the potatoes will form. From this time, continue hilling up in small increments as weeds emerge between the rows. Do this until the vines start falling over, after which further hilling is not possible.

If you hilled enough, there will be no potatoes forming that are exposed to the light, so there will be no green potatoes to throw away. From this point on, hand-pull any weeds appearing among the vines. There should be next to none. If the crop has grown well it should be nearly impossible to walk between rows on three-foot (90 centimeter) centers without damaging the vines. From this point, it is best to keep your soil-compacting feet out of the plot anyway.

Pests and diseases. All this hilling gives you a good opportunity to check for and control potato beetles if you're in the eastern United States.

Every expert says not to lime potatoes because it causes scab. I have not noticed any difference, lime or not, but I am providing the obligatory warning here. I usually remember to keep the lime out of the bucket of COF I use to fertilize my potatoes.

There are numerous soil diseases that affect potatoes. It is a good idea to grow spuds on a new piece of ground that hasn't seen them nor any other solanums (tomatoes, peppers, eggplant) for at least three years.

Varieties. I can't predict the best varieties for your locality. However, I can tell you that there are early, midseason, and late-maturing ones. To understand how this is, you need to know how the vine works. First the plant grows leaves.

All surplus food made by these leaves is used to make more leaves. Then the plant begins to bloom; at this point, tubers appear as little below-ground nodules along the stems immediately above the seed piece. The nodules begin to fatten into potatoes. At this stage, growth of new leaves slows. A few weeks later, blooming stops. When flowering ceases, production of new tubers, new vines, and new leaves also ceases. The existing leaves pump out sugar and other nutrients that are translocated into the already formed tubers and stored there. This goes on until the vine dies off. As it shrivels, all the nutrients the vine contains are also translocated into the maturing tubers. At this time the potato skins toughen up as they prepare to endure winter.

Early varieties grow for less time before beginning to bloom. They make fewer leaves on shorter vines and thus are done sooner. They also yield less. Late varieties grow on longer before the bloom starts, and they bloom for a longer period. Lates yield more. In climates where there is a long growing season, it is often wise to use the latest of late varieties for the main storage crop, or even to delay planting the main crop for a month after the earlies are sown, because if the main crop vines die back just before the frosts come, their potatoes will have less tendency to resprout prematurely. When lates are finally dug, the weather will have become cool, leading to much longer storage. I always grow a small patch of early potatoes for summer use, but most of the crop will be late varieties.

Potatoes don't like scorchingly hot weather, especially when it is also humid. For that reason, they are usually grown as a spring crop where summers are long, hot and humid, and are dug early.

Harvest and storage. As soon as the earliest sowing of the earliest variety is in bloom, you can dig a plant for new potatoes. They'll grow one size every few days as you dig your way down the row. Some people try to gently tickle out the odd new potato without damaging growing plants, but I find this usually causes more loss than it is worth. I just dig an entire plant whenever I need new potatoes.

Do the main harvest when the vines have browned off, indicating that the skins are tough and won't rub off easily. Dig carefully so as not to cut potatoes. Cuts do not heal well enough for long storage; any cut potato must be eaten within weeks. Do not bruise the potatoes; handle them gently at all times. Using virus-certified seed, I harvest about 25 pounds (11 kilograms) of potatoes for every pound (0.5 kilogram) of seed sown. My yield would be half that with diseased seed. Maybe even less.

Ideal storage conditions are quite humid and a stable 40°F (4°C). Colder than that and the spuds become sweet as the starch converts to sugar. If the temperature goes below freezing, potatoes will be ruined. Warmer than 40°F increases their tendency to resprout. I recommend you study Mike and Nancy Bubels' *Root Cellaring* for suggestions of less-formal ways to store things over the winter than in a cellar. I live in a maritime climate and put my spuds in large cardboard boxes that are stacked in a tightly built, unheated outbuilding. I cover the boxes with a few old woollen blankets to make sure no light gets in and also to stabilize the temperature. Given this small attention, they last about five months before sprouting starts, and even then they are quite useable for another five or six weeks if I rub off the sprouts, by which time the first new potatoes are on the horizon.

Saving seed. Don't save seed unless you absolutely have no choice. The odd aphid passing through infects the vines with assorted virus diseases. These do not kill the plant, but they do reduce its vitality and lower yield — a lot. There are over 20 such diseases and they are transmitted from year to year in the seed itself. The more years gardeners plant from their own seeds, the more viruses the potatoes will contain and the lower their yield will become. It is possible to harvest more than double the yield by sowing seeds certified to be disease free. Such seed is not costly compared to the result it provides.

If you do save your own small spuds to plant the next year, do not do this for more than one or two years before starting anew with certified seed. And do not try to use supermarket potatoes for seed, both because they are not certified and also because, unless they are genuinely organically grown, they have been treated with anti-sprouting chemicals (which gives me yet another reason, beyond their lousy flavor, not to eat commercial spuds). Seed certified as disease-free begins with tissue-culture, laboratory-grown mini-tubers that is grown for a few generations in high-elevation or chilly, isolated places free of aphids. This has nothing to do with being certified as organically grown.

Sweet potatoes

Growing details. Loamy to sandy soils are essential for really good results. It is nearly impossible to lighten up a clay soil enough to grow the finest sweet potatoes because too much manure or compost will cause the quality of the potatoes to suffer.

The root system sprawls to match the vines, so spread a quarter- to a half-inch-thick (6- to 12-millimeter) layer of well-rotted manure or finished compost over their entire growing area, then dig the entire area. Sweet potatoes also grow well if you dig in a thick stand of an overwintered legume green manure. Form wide raised rows four feet (120 centimeters) apart on center. Because good drainage after heavy rains is essential, raise their beds about six to eight inches (15 to 20 centimeters) above narrow footpaths between them. Once their beds are formed, set either vine cuttings or rooted shoots 15 inches (40 centimeters) apart atop these ridges. They may also be arranged with three seedlings planted in a moderately fertile hill, the hills on four-foot (120-centimeter) centers.

In the right soil type, given only moderate fertility, and grown where summers are long and nights are warm, each plant can produce 8 to 12 good-sized potatoes. When grown in the northern United States, the yield may drop to as low as about one pound (450 grams) per row foot (30 centimeters). You can save planting stock from the previous year's crop or buy ready-to-plant shoots. If the seedling raiser is reputable, buying may be the best option because sweet potatoes accumulate virus diseases like Irish potatoes do. They also have a tendency to mutate. Commercially raised shoots or seedling are usually only a few generations away from pure, virus-free tissue-culture clones. Statistics show that virus-free starts increase yields by as much as a third.

To start only a few vines, in a warm place suspend a sweet potato on toothpicks in a container and cover half of it with water. You can grow larger quantities by placing several presprouted (or conditioned) sweet potatoes on a bed of sand, covering them with a two-inch (five-centimeter) layer of moist sandy soil, and keeping that soil between 70°F and 75°F (21°C to 24°C). Presprouting (what is called "conditioning" in the commercial trade) consists of holding the roots at about 85°F (29°C) and high humidity for a few weeks until shoots start to develop. You can easily accomplish this at home by putting a few roots into a germination box (for heat) that also holds a large open pan of water (for humidity). Don't allow the presprouting chamber to become so humid that the roots are actually wet; this may cause mold or fungus to form on them.

In the deep south of the United States, outdoor nursery beds are used. The soil will naturally be at the right temperature to sprout the roots about a month to six weeks before the correct time to set the shoots out. The time to start the nursery is when the night temperature stops falling below 60°F (15.5°C). If you're farther from the equator, you may have to bed the roots in a greenhouse or indoors, putting a soil- or sand-filled box in a bright sunny window in a warm room. The sprouts will be the optimum size for transplanting when they have grown 10 to 12 inches (25 to 30 centimeters) long and have five or six leaves and a stout stem. Cut off the sprouts two inches (five centimeters) above the soil with a sharp knife. Because they're unrooted, put them into soil almost horizontally, about two inches deep, making sure to allow two leaves to be above ground. Take care not to damage the shoot's terminal bud. Immediately after you set them in their bed, water well.

The light soils that sweet potatoes prefer dry out rapidly, so as the plants grow, keep the soil moist, if possible, but never soaked. This encourages better root development. Keep the area well-weeded before the vines run too

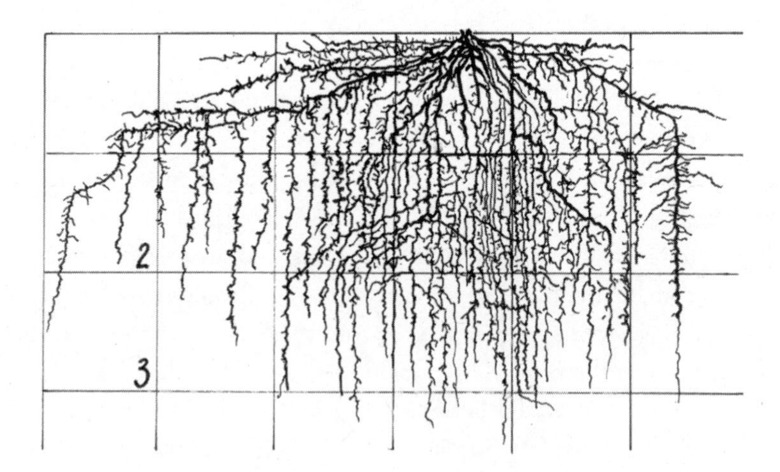

Figure 9.4: *A sweet-potato plant at midsummer. With rows four feet (120 centimeters) on center, the vines from one row are reaching the plants in the next — as are the roots. No potatoes have formed yet. At harvest the vines and their roots will both have reached 14 feet (425 centimeters) at their greatest extent and be working to a depth of about 4½ feet (140 centimeters). Do not crowd sweet potatoes! (Figure 69 in RDVC)*

much. It's best to do this by shallow scraping with a hoe, much as suggested for Irish potatoes. Gradually make a low hill around the stem of the vine by scraping up weeds and soil against it. Louisiana State University agricultural extension service says hilling reduces insect problems.

Under good conditions, within two months of being set out the vines should entirely cover the spaces between rows. From then on do a bit of hand-weeding (and careful stepping) if you want the best results.

Pests and diseases. Commercial producers are distressed when any fraction of their crop fails to appear marketable, so pests and diseases are mainly a cosmetic worry to them. They are usually not a serious matter to the home gardener, except that there may be quarantines on the movement of potatoes and planting stock in states with major sweet-potato production. Be warned: at the time of writing, Louisiana, Alabama, Georgia, and North Carolina have restrictions. Check with your local agricultural extension branch for information on your area.

The gardener's most effective weapon to prevent problems is crop rotation. Rotation also helps simplify weeding. After at least two years of growing some other cleanly cultivated crops, the area will be relatively free of weed seeds and ready for sweet potatoes. A weedy sweet-potato crop is a stressed crop, a non-productive crop, and a crop attractive to pests.

Do not repeat sweet potatoes on the same beds without a rest into other crops lasting at least three or four years. During that break you should not allow any of its pernicious relatives — like bindweed or morning glory — to grow. Finally, carefully clean up all vines, dig out all accessible root materials, and promptly hot-compost (or burn) them to greatly reduce pests for the next year. Due to companionate effects, legumes following sweet potatoes won't grow well. It's best to follow sweet potatoes with a brassica cover crop.

Varieties. In addition to the usual orange- or red-fleshed soft, sweet types, there are dry-fleshed white ones. These are quite popular with people from the Caribbean. I know of but one variety requiring less than 120 to 150 warm days and warm nights — Georgia Jet needs only 90.

Harvest, curing, and storage. Harvesting and storing seem to be the most rigorous aspect of growing sweet potatoes, especially growing early-maturing varieties in short-season climates. Under warm conditions the first harvest can begin about 90 to 120 days after setting out the shoots, but the potatoes will continue to swell in size for some time after they are first large enough to eat. Most of the sizing up happens rapidly during the last few weeks of growth, so watch closely. You want them sizeable but not overly large because eating quality suffers when they're too big. (This is one reason not to make their soil too fertile.) There will come a point when all the sweet potatoes should be dug promptly.

It is important to harvest gently to minimize "skinning," the scraping off of skin. Skinned potatoes don't keep well. If the soil is quite dry at digging time, a light irrigation is helpful to soften it and reduce damage. However, if the earth is soggy when digging, the potatoes may crack open after harvest; they may even rot in the ground.

Select the roots you will use to make the next year's seedlings while digging. These should be 1½ to 2½ inches (four to six centimeters) in diameter with smooth skins, a pleasant appearance, and no sign of insect damage or disease. They should also be taken from hills that made an abundant "nest" of potatoes.

Do not allow just-harvested potatoes to sit in the sun for more than one hour or they may scald, reducing their storage potential. And do not, if at all possible, harvest after frost. If you should be caught by a surprise frost, then harvest within days or the roots may begin to rot. Sweet potatoes still in the earth are badly damaged if they experience soil temperatures below 50°F (10°C).

After harvesting, the potatoes should be cured to heal any scrapes or injuries to their skins. Ideal conditions for this are a stable 85°F to 90°F (29°C to 32°C), humidity around 90 percent, and reasonable ventilation. Commercial growers have special rooms designed to maintain these conditions. North Carolina State University agricultural extension service advises the home gardener to use this simpler method: Once the roots have been removed from the garden, spread them out to dry for several hours away from direct sunlight. Once dry, put them in newspaper-lined boxes and leave them in a dry, ventilated area for two weeks for curing. When they are cured, store them in a cool, dry place (50°F to 55°F/10°C to 13°C). Wait a month until they've sweetened up and you'll be ready to start cooking them. Most varieties of newly harvested potatoes do not taste sweet. Their flavor develops over time. In storage, beware of cold; if the roots get colder than 50°F (10°C) for even a few days, the core gets hard and loses quality. On the other hand, if they're kept too warm for extended periods, they may shrivel, become stringy and pithy, and/or sprout prematurely.

Curing and storing sweet potatoes so that they'll last the entire winter is more difficult than growing this easy crop, especially where weather conditions are cool at harvest time. If you master the art of keeping them until the next planting season, you may wish to start more than a few dozen seedlings in the spring because then this crop could become your nutritious basic staple, every bit the equal of the Irish potato.

Tomatoes, eggplant (aubergine), and peppers (capsicum or chilli)

Tomatoes, peppers, and eggplant are close relatives. Many varieties can be perennial where there is no frost. All are aggressive growers in suitable weather conditions, responding to fertilization by expanding to the limit of their moisture supply and rooting room. Gardening magazines occasionally show a photo of one trellised tomato plant covering the entire sun-facing wall of a house, and I was once introduced to a five-year-old Fijian eggplant bush that was five feet (150 centimeters) tall and six feet (180 centimeters) in diameter. Once a year it was trimmed back by half and its bed was mulched with a few gallons of chicken manure. I suggest new gardeners learn to grow tomatoes first; once they are mastered, peppers and eggplants will seem easier.

Growing details. Tomatoes are not frost-hardy; most varieties need 100 to 120 growing days from emerging to first ripe fruit. Where there are fewer than 150 frost-free days, gardeners must get at least a 50-day head start by using transplants. Those in warm climates may direct-seed tomatoes; usually these folks still use transplants to obtain extra production time. My advice: If your frost-free growing seasons exceeds 150 days, grow or buy only two or three early-maturing transplants of a bush determinate variety, enough to supply the table early in the season. Then directly seed a few more spots at the same time the seedlings are set out, and make these indeterminate varieties. This makes life simple. It also frees you from any temptation to buy a lot of tomato plants.

To direct-seed tomatoes immediately after there is no further frost danger, use hills spaced on at least four-foot (120-centimeter) centers. Gently press the soil back down to restore capillarity, make a thumbprint in the center of that mound about half an inch (1.25 centimeters) deep, put five or six seeds in that depression, cover with loose soil, and, if it is hot and sunny, water that spot every day for about a week. Progressively thin the seedlings to a single plant per hill.

You should lift tomato vines off the earth, freeing the fruit from damage by insects or rotting. You can stake up indeterminates, trellis them, guide their vines up hanging strings and wires, grow them in tall wire cages, etc. The best way I know to prune or train indeterminates (and there seem to be as many ways to do this as there are gardeners) is to follow their own growth pattern. Leaves, and the new vines emerging from their notches, form on the stem in threes. First, two weak side branches appear; the third one is a stronger side branch, and this pattern then repeats. During the first few months of the plant's growth, remove all the weak side branches as soon as they appear and allow the third strong ones to grow. This is easily accomplished by pinching

The terms "indeterminate" and "determinate" refer to different growth patterns. All tomato vines grow by having one side branch emerge from each leaf notch. This side shoot will grow as freely as the main vine unless pinched off. Determinate vines grow only a few leaves (usually three) and then stop. The side branches continue the growth for three leaves and then stop, and so forth. Determinates tend to be tidy, compact plants.

Indeterminate varieties make a vine that keeps growing indefinitely from its end; its side branches also grow indefinitely. The result is that the vines on indeterminates are much longer, and usually the length of stem between each leaf is also longer, so they are lanky and spread aggressively.

Both sorts will keep producing new fruit and covering new ground until frost — as long as the plant can find unoccupied soil in which to make roots. If the root zone gets crowded, they become stressed, their production slows a lot, and often the vine becomes diseased or will be attacked by insects. The longer your frost-free growing season is, the more growing space you should give tomatoes.

The root systems of determinate varieties closely match their compact above-ground growth. Determinates set more fruit sooner, ripen it sooner, and yield more heavily for a shorter time. Because of their growth habit, determinates are not staked up or trained and are bred so as to hold much of their fruit slightly above the soil. Short-season gardeners should probably only grow these types because they're inevitably earliest. Generally the flavor of determinates is second-rate because the vine carries more weight of fruit in proportion to the amount of leaf area it makes to fill those fruit with taste and nutrition. ■

off the unwanted side shoots with your thumbnail. Thin both the main branch and also all the side branches that you allow to remain. Continue pruning this way until the plant has formed enough branches to suit whatever mechanics you have constructed to hold them up. After that, remove all side branches as soon as they appear.

An old-fashioned way to lift determinate vines enough to prevent most damage is to spread a few inches of dry brush to cover where they are to grow and let the vines grow over it.

Pests and diseases. There are many diseases; most are a problem only in commercial fields. If the roots have room to expand, the weather is favorable, and the soil is reasonably fertile, the vine usually won't become sick.

Fruit worms are the same larvae as the ones that eat corn and may be controlled by spraying Bt (see the section on the corn earworm in Chapter 8).

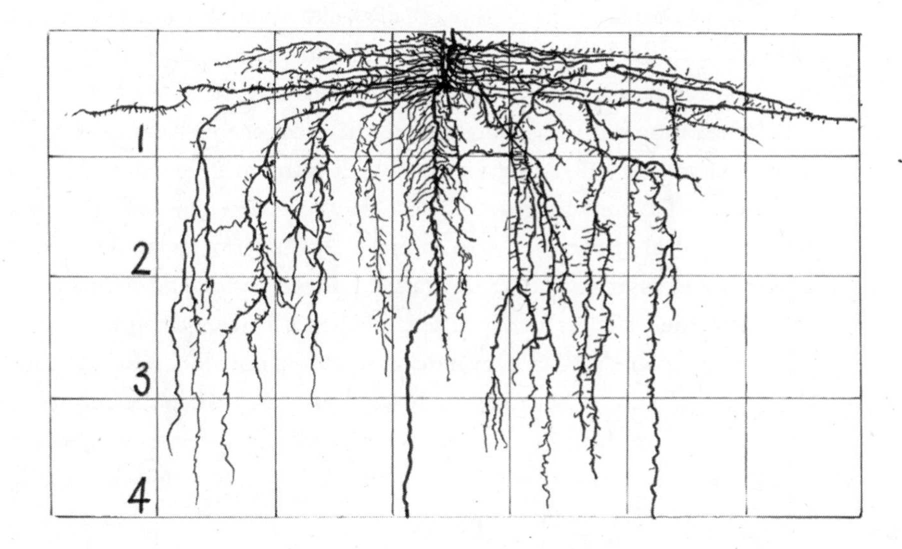

Figure 9.5: *An indeterminate tomato plant, spaced four feet by four feet (120 by 120 centimeters), shown two months after transplanting. The tops aren't bumping yet, but the root systems are. At the time of the first frost, this plant had roots extending five feet (150 centimeters) from the center, and the entire system had thickly penetrated an eight-foot-diameter (245-centimeter) circle to a depth of 42 inches (105 centimeters). Clearly, tomatoes need a lot of growing room to do their best. (Figure 73 in RDVC)*

Hornworms eat leaves; they also die after eating a bit of Bt. Blossom-end rot will stop being a problem on most varieties once your subsoil offers sufficient calcium; it can take a few years of light applications of lime for enough calcium to build up in the subsoil. Have faith: end rot will fade away.

Varieties. There are a great many varieties of tomatoes. Hybrid varieties are only slightly more vigorous. Their big advantage is that they can be resistant to more diseases at once, which is important to commercial growers who run down their soil's organic matter content, overcrowd the plants, and don't rotate out of tomatoes often enough. Some heirlooms have superior flavor but may not be well-adapted to cool or dry climates, and they rarely carry any disease resistance at all. I suggest that you only experiment with heirlooms. Try a new one every year, but mainly grow the tomato varieties proven to work in your area. The best flavor is found in slicing (firm-fleshed, not watery) varieties that require warm humid nights. These are often called "beefsteak." However, in maritime climates these classics usually fail to ripen, which is another reason to check with the local experts about locally adapted varieties. Some varieties may deal better with diseases present in your area, too.

Indeterminate cherry tomatoes are the most aggressively growing of all types and are best suited to dry gardening. If you're making sauce or paste, use varieties bred to contain less moisture. They cook down in half the time, saving a great deal of energy. These sorts are also superior for drying. Finally, a relatively new sort has come out called Longkeeper. It and its competitors ripen extremely slowly; after you pick it when it's green, the fruit develops a waxy tough skin that holds in moisture, lasting a lot longer while ripening slowly in the pantry. They taste pretty good, especially when the only comparison comes from the supermarket. I suspect that Longkeeper was bred from an old Burpee classic called Golden Jubilee, a late-maturing yellow beefsteak type that has, in my opinion, the best flavor of any tomato. I always grow one Golden Jubilee plant even though it is too late for my climate and even though I only get a few fully ripe fruit toward the end of summer unless the year has proved unusually warm. Still, the full-sized green ones are among our best in-the-house ripeners.

Peppers and eggplant are less-vigorous close relatives of the tomato. They too are self-pollinating, but have a slight tendency to outcross. If you are saving seed, give varieties about 20 feet (six meters) of isolation. It's especially important to isolate hot peppers from sweet ones. The seed-saving procedure is identical; let the fruit get dead-ripe first.

Peppers are sometimes direct-seeded where the growing season exceeds 150 days (I know someone in Maryland who does that); I wouldn't try it with eggplant unless the growing season exceeded 180 days. In short-season areas or in maritime climates, growing hybrid peppers is highly advantageous. In climates where the summers aren't hot, if you don't have a greenhouse, hybrid eggplant may be essential, and the best of the lot in Cascadian trials is Dusky Hybrid. In chilly areas, it may be to your advantage to grow peppers and eggplant on top of a black plastic mulch that covers their entire wide raised bed. The mulch warms the soil a few degrees and also increases the nighttime air temperature a few degrees. This little rise in temperature makes all the difference. It is not necessary to use drip or trickle irrigation beneath such mulch in the garden. Simply make the bed perfectly flat. Then lay and anchor the mulch and set transplants through small "X" slits. When overhead watering or when it rains, puddles will form in slight dips and depressions. Poke small holes in every one of these spots to let the water flow through.

Harvest and storage. I pick my scarce first fruits when they're light orange because I hate to see any of them damaged by slugs or woodlice. After trellised indeterminates are ripening higher off the earth, there is less danger of damage (and many more fruit to spare), so I let them develop a completely vine-ripened flavor. At the end of the season I bring all full-sized green tomatoes into the house and keep them in airy baskets in a cool part of the house to ripen over the next few months.

To improve the performance of determinate varieties, about eight weeks before the first expected frost, pinch off about half the flower clusters as they appear. This lightens the fruit load and enhances size and flavor. About four weeks before the first expected frost, pinch off over three quarters of all the flower clusters as they appear.

Saving seed. There is little advantage to growing hybrid tomatoes, so you may as well grow open-pollinated sorts and save your own seed. Tomatoes (and peppers and eggplant) are self-pollinating, and you can save seed from a single plant. Let the fruit used for seed extraction get completely ripe on the vine; it'll sprout better and store longer if you do. Then bring these dead-ripe fruit into the house, keep them in a warm place, and let them ripen even further for another few days.

Cut the tomatoes in half crosswise, exposing all the cells inside. With your finger, scoop out the juicy (seedy) pulp into a bowl. Put that watery mixture into a small jar that holds double the volume you are putting into it. Allow it to stand on the counter for a few days in a warm room. It'll ferment, fizz, and foam, and white mold may grow on top. That's fine. The acids caused by fermentation dissolve the sprouting inhibitors within the seed, making it much more certain to germinate. Swill the liquid around in the jar once a day to help settle the seeds out of the froth on top. Within three to five days most of the seeds will have settled on the bottom and most of the solids will be floating on the top. Now gently fill the jar with water, allow the seeds to resettle to the bottom, and gently pour most of that water off without losing many seeds. Refill and repeat this until the water is clear and the solids are gone. The few seeds you will lose doing this are to your good; these float to the top because they are lightweight and unripe.

Pour the contents of the jar through a tea strainer to catch the seeds. Rinse the jar a few times to get them all out. Then rinse the seeds in the strainer under running water until clean. Dump the seeds on top of several thicknesses of newspaper and let them dry for a few days before you put them away in a paper envelope. With most varieties, a few ripe tomatoes will provide enough seeds for the entire neighborhood.

Winter squash (pumpkin), zucchini, melons, and cucumber

Cucurbits are so similar that if you have known one, you know them all. The easiest one to start with is squash, because it is the most vigorous, most tolerant of chilling, and most tolerant of heavy soils.

Growing details. The key to success with cucurbits is to wait until the soil warms up enough before sowing. And you should almost always directly seed; they don't transplant well. Whether while sprouting or when growing on, cucumbers need more warmth than squash, and melons (especially watermelons) need more than cucumbers. Sow squash and pumpkins after soil has warmed to 60°F (15°C) and ideally sow during a week of sunny weather. After squash seedlings are up and have begun to grow well, sow cucumbers. As soon as cucumber seedlings are up and beginning to grow well, sow melons. This timing method matches the steadily increasing temperature of soil in spring.

Any time the soil turns chilly and damp while cucurbit seeds are sprouting, they may die. Cucurbits are more sensitive to chill and damp before coming up than after. Should any young seedlings (or seed not yet emerged) experience a spell of chilly or rainy weather, remedy the situation using this method: As soon as the weather settles, resow at another spot in the same hill. If both sowings end up growing well, progressively thin both and finally choose the plant that grows the best. Sometimes in a chilly damp spring, a sowing made a week or ten days later will outgrow an earlier one that barely emerged, shivering and shaken.

For the same reason, you should be reluctant to water sprouting cucurbit seeds. It's better to chit them first, plant them into moist warm soil, and then not water them at all until they emerge.

Cucurbits grow fast in full sun and in fertile soil (most fruiting plants do). Since the mainly shallow root system of all cucurbits is at least as extensive as their tops are (with bush summer squash, the roots may spread significantly more than the leaves do), the entire area their vines will ultimately cover should be made fertile; additionally, to get the seedling off and growing fast, sow seeds in a hill with extra fertility beneath it.

On hot sunny afternoons, gardeners often shrug off temporary wilting of cucurbits as unimportant. The attitude is: Inevitably the plant recovers, don't worry. This is not correct; any wilting is a big stress and greatly reduces plant health and overall yield. Cucurbits don't wilt on hot days when only one plant grows in each hill and the plant does not share its root zone. So where squash borer is not a problem (and the insect is rarely a problem with cucumber and melons), do not grow two or three plants per hill. Start three but thin to one by the time the vines start to run.

Watermelons are intolerant of heavy soils.

Pests and diseases. See the discussion in Chapter 8.

Varieties. Hybrid squash varieties now dominate in seed catalogs. Consequently, OP summer squash varieties have degenerated, excepting the old Yellow Crookneck (vining), which is by far the best-tasting of all summer squash and resists pests better, too. Hybrid winter squash may outyield the classics by half again, but they don't taste any better, and for most people the production of one or two good hills of ordinary varieties is a gracious plenty. Incidentally, if you want a really sweet and long-keeping winter squash, try the

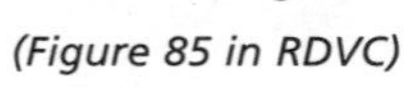

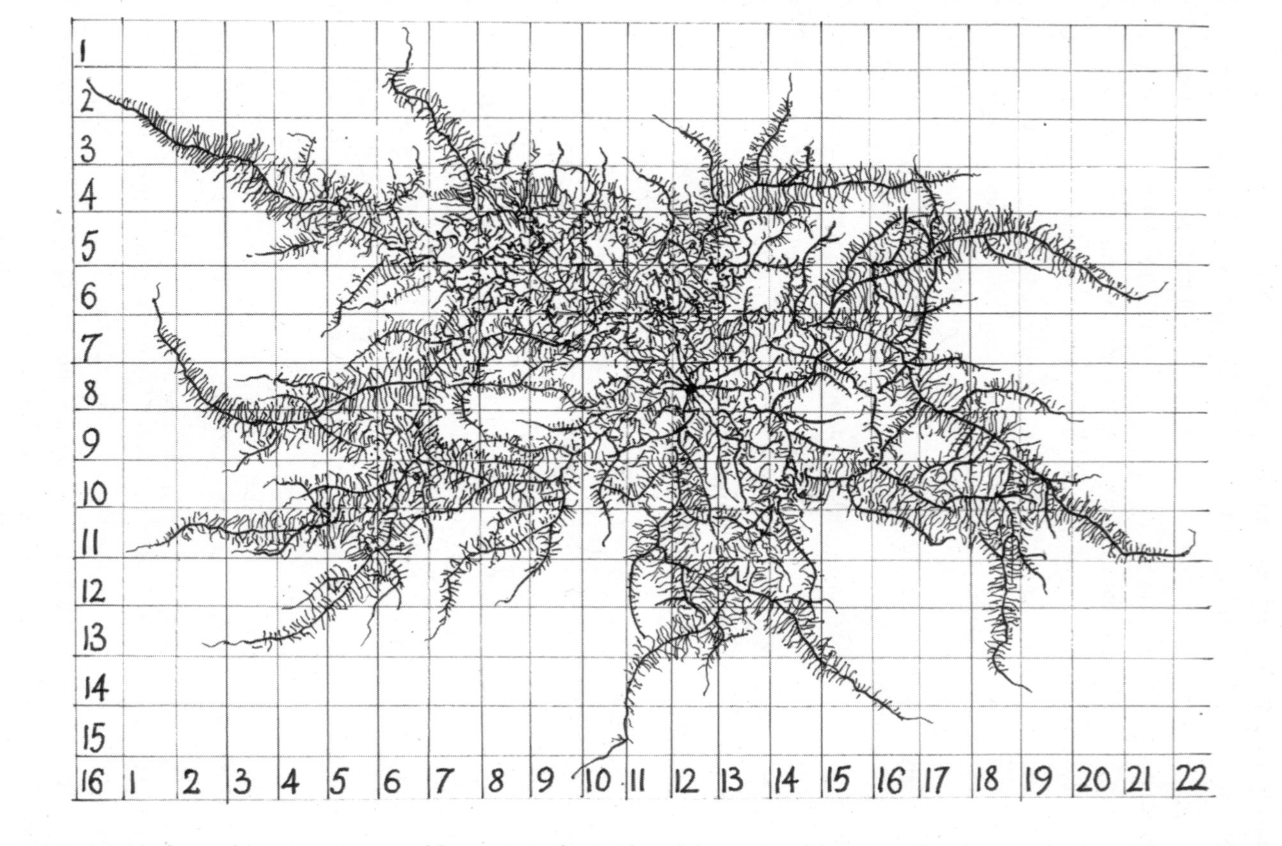

Figure 9.6: *A typical cucurbit root system — this one is Rocky Ford cantaloupe in midsummer. The drawing is viewed from above. The extensive root system is shallow, rarely going down more than two feet (60 centimeters). Winter squash, a more vigorous plant, makes roots covering half again more area. With cucurbits, you can assume that the roots are always at least as extensive as the vines are. (Figure 85 in RDVC)*

genuine heirloom Sweet Meat (from Harris Seeds or Territorial Seeds) or one of the somewhat shorter-storing delicata types (including Sweet Dumpling). Acorns don't keep well at all. Australians have an especially long-keeping pumpkin named Queensland Blue (what North Americans call winter squash, people Down Under call "pumpkins") that makes the best pumpkin soup you ever tasted. Johnny's Selected Seeds sells an heirloom variant of this from the state of Western Australia called Jarrahdale.

Of cucumbers, the old sprawling classic "apple" or "lemon" is best adapted to lower levels of fertility and soil moisture. Hybrid cucumber seed is affordable; hybrid melon seed isn't. However, in the cool nights of a maritime climate, only the earliest of hybrid melons will produce anything, and only when grown atop a wide sheet of black plastic. I spread a piece about six feet (two meters) wide, anchor it on the edges, and put one vine every four feet down the row. That's the only way I can harvest ripe melons in a maritime climate.

Harvest and storage. Everybody Else says zucchini and other summer squash stop yielding much after a month or so, but it doesn't have to be that way. If you give them growing room, the plants will produce more growing branches, more flowers will form, and the yield will steadily increase until the weather turns against the crop. So don't crowd them. I have grown bush varieties in hills on five-foot (150-centimeter) centers with great results. ("Bush" squash are really just vines with short spaces along their stem.)

Remove all overlooked oversized fruit from summer squash and cucumber vines because the burden of forming seed reduces formation of new fruit.

Winter squash don't taste great until they are fully ripe. They have reached that state when the stem attaching the fruit to the vine has shriveled and become brittle, revealing that it is no longer passing vascular fluid. Most winter squash in the species *Cucurbita pepo* (acorn, delicata) don't store as long as those in *C. maxima* (Hubbard, buttercup) or *C. moschata* (butternut). Ideal storage is 60°F (15.5°C) with low humidity and good air circulation. Warmer and dry is better than cool and damp. Storage time has a lot to do with variety. I've also found that leaving winter squash outside to experience more than the first light touch of frost greatly lowers their storage potential. Curing also helps lengthen storage. Bring them inside where it is warm and dry for two weeks; this toughens their skin. We heap ours up in the dining room. Then, cured, they go to a cooler dry place.

Cantaloupe, honeydew, and similar melons are ripe when the vine slips off the fruit with only slight pressure. They do not ripen after being picked, which is why supermarket melons, inevitably harvested when unripe (and still hard enough to pack in a box and ship), are inferior. To determine when to cut watermelons from the vine, you must thump them and knowledgeably listen to the sound they make. I can't explain this talent in words. It must be developed with practice.

Saving seed. Cucurbits are pollinated by bees. Isolation of at least half a mile (800 meters) will prevent enough crossing for home-garden purposes. There are three commonly grown squash species: *Cucurbita pepo*, *C. maxima*, and *C. moschata*. These species do not cross, but every variety within one species does cross with every other. All summer squash, delicata types, acorn squash, and most jack-o'-lanterns are *C. pepo*. The large winter squash varieties are usually maxima.

All but a few cucumber varieties are within the same species and cross.

Canteloupes (rockmelons) don't cross with honeydew types. Some specialty melons are unique species. Check the seed catalog for their Latin names and assume different species won't cross-pollinate. The seed is ripe when the fruit has become completely ripe (slips the vine).

Practically speaking, all this interesting information is irrelevant to the home gardener; when it comes to growing your own seed for these species, I strongly suggest you don't try for more than one generation unless the seed-producing population exceeds 25 plants. Saving seed from only a few cucurbit plants leads to inbreeding depression of vigor; within a few generations the seed will barely sprout or grow. Few gardeners grow 25 winter squash vines! However, the seed is long-lasting and the first generation's seed collected from just a few fruit will supply you and the whole neighborhood for the next ten years if properly stored. My suggestion is that at least every other generation you buy a packet of new stock.

Some maxima varieties have seeds that taste good, but some don't. Sweet Meat's seeds, for example, make great munching. You may be discarding the best of this vegetable's nutrition if you don't extract, dry, and eat its seeds.

Beets (beetroot) and Swiss chard (silverbeet)

These quite different vegetables derived from the same wild plant, whose Latin name, Beta, became "beet" in English. The vegetable "beet" was bred for the succulent sweet root; thinnings grown to the stage of developing baby-sized beets are often cooked, tops and all. Swiss chard was selected to emphasize large tender leaves, but dig one up and you'll see a poorly formed beetroot. Beta, with a huge reservoir for moisture storage in the top of its root and a deeply adventuring root system, is good at handling long-lasting dry spells. If there is nutrition accessible in the subsoil, Beta does not need hugely fertile topsoil.

Growing details. Technically, beet seeds are fruits; each usually produces several seedlings. The commercially produced seeds can be stored a long time; six years is normal and ten isn't unusual. So if you get some beet or chard seed that sprouts poorly, take it as an insult; it had to have been really sad old stuff. To get well-formed beets, careful thinning is essential. Shortly after germination, reduce the thickest of the clumps, but do so cautiously as there will usually be a fair number of mysterious seedling disappearances. Postpone the final, precise thinning until the plants are about four inches (10 centimeters) tall. Then thin them to whatever spacing you'll want the mature beets to be growing to.

Varieties. Where winter consists of only frosts or chilly weather, two historic and virtually identical varieties, Lutz or Winterkeeper, are bred to make enormous roots that hold for months. Cylinder varieties are bred for canneries that want to slice rounds and also need quick-cooking vegetables; they have little fiber. White beets (sugar beet crosses) have a sweeter flavor.

Harvest, storage, and use. Most varieties of beet will continue enlarging without becoming woody or tasteless if only they have room to grow. Harvesting by pulling every second beet in the row helps the patch remain in better eating condition. Thinning to a wider spacing at four inches (10 centimeters) tall and using wider between-row spacings helps even more. I once dry-gardened delicious beets spaced one foot (30 centimeters) apart in rows four feet (120 centimeters) apart. After five entirely rainless months, each root was nearly the size of a volleyball and still delicious.

In mild climates, beets may overwinter in their bed. In Cascadia — where in the rare year a short spell of freezing weather may ice their crowns, killing them — burying the crowns under a few inches of soil is sufficient protection. Beets root-cellar well in a barrel of moist sand.

Because this crop is so well-suited to being a vegetable staff of life, I thought I'd suggest a few ways you can use beets that most people do not know about.

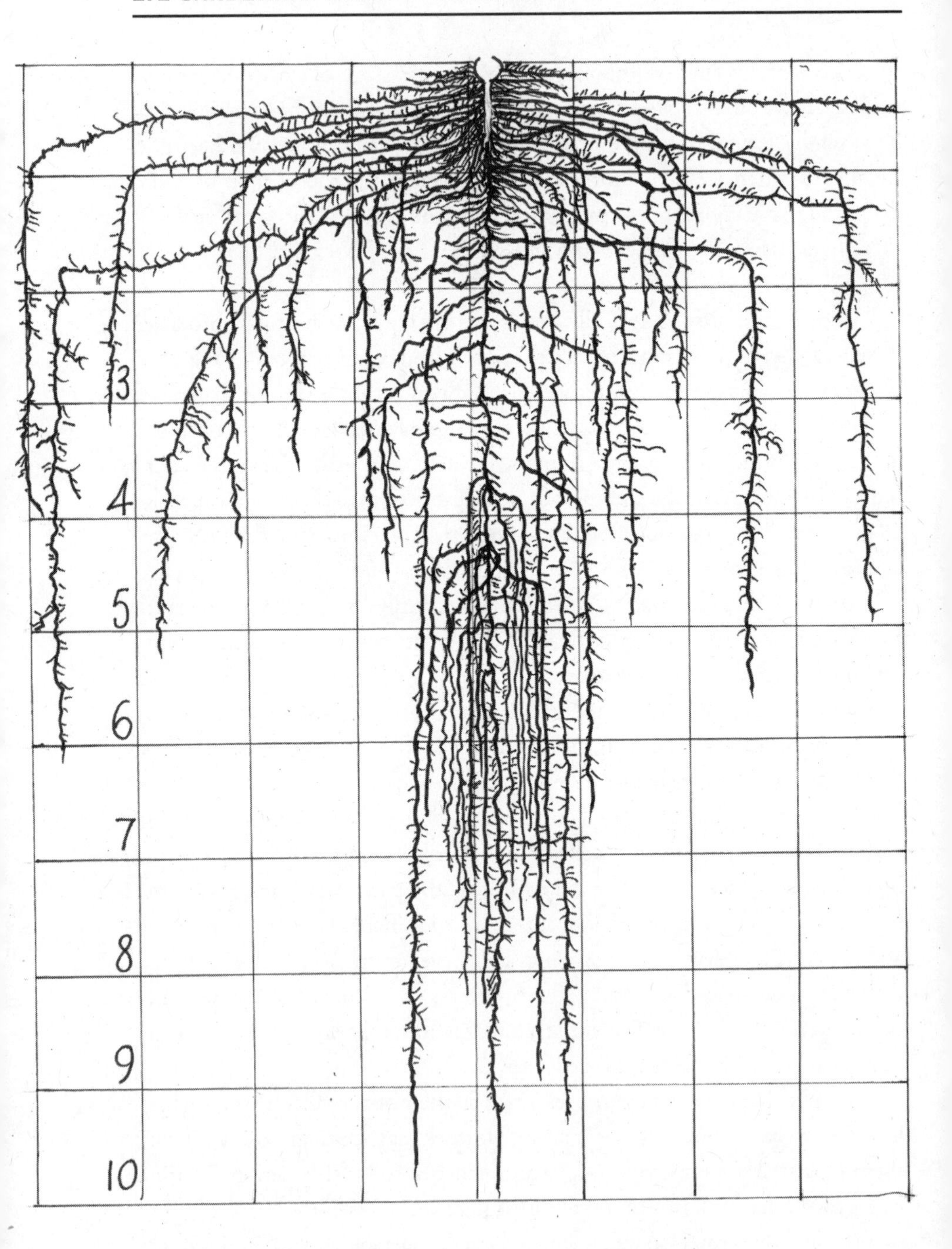
3
4
5
6
7
8
9
10

Beets may be baked like potatoes and are delicious. This method of preparation retains more of their nutrition than boiling does.

Raw, grated-beet salad can be delicious in areas where the subsoil offers balanced fertility; given proper nutrition, the beetroot will not have any of that back-of-throat-rasping sensation many associate with raw beet. Mix grated beet with bits of navel orange and minced mild onion; dress with a bit of lemon juice and black pepper.

Saving seed. Beets must overwinter before making seed the next spring. Wind-pollinated, they need at least a quarter mile (400 meters) of isolation for rough purity. Seedmaking plants occupy a lot of space and inevitably make lots of seed. In early spring you should replant the roots to the depth of their crown at least one foot (30 centimeters) apart, in rows three to four feet (90 to 120 centimeters) apart. In midsummer, when much of the seed is dry on the stalk, harvest the whole plants and let them finish drying to a crisp under cover on a tarp. Then rub the seeds off their stalks and winnow to clean away the dust and chaff. To maintain a variety through more than one generation, include at least 25 plants in the gene pool. Four to six plants will make several pounds of seed that will last a decade if it has not been rained on or irrigated during the drying-down period. This is hard to achieve in areas with summer rain, which is why commercial beet seed is grown in Cascadia, where the summer is reliably dry.

To make chard (silverbeet) seed where the snow flies, dig half a dozen plants with as much root as you can get up, cut off their larger leaves, plant them in large tubs or beds of earth in the root cellar over the winter, and then transplant them back outside in spring. In moderate winters it might work to hill up soil around and actually over the plants, and then uncover them in early spring. Otherwise they are like beets and will cross with beets.

Figure 9.7 *(at left): A beet root system about 110 days after sowing seeds. Clearly a single plant is designed to make use of all the moisture and nutrition from a cylinder of soil about two feet (60 centimeters) across and seven feet (215 centimeters) deep. Thus, in dry gardening, a useful spacing for this drought-tolerant crop would be about one foot (30 centimeters) apart in rows about four feet (120 centimeters) apart. (Figure 21 in RDVC)*

Sweet corn

Growing details. I prefer growing corn with a bit more elbow room than most because the plant has a natural tendency to tiller, meaning it will put up additional ear-bearing stalks if there is enough growing room. And even if a plant doesn't tiller (tillering is a trait modern breeders try to eliminate because ears forming on secondary stalks are smaller), having a bit more soil to access will protect it against drought and also make the main ears get a bit bigger.

Where soil moisture is not a problem, each plant should exclusively control at least 2¼ square feet (2,000 square centimeters). Eighteen inches (45 centimeters) on center or nine inches (23 centimeters) apart in rows 36 inches (90 centimeters) apart, or eight inches (20 centimeters) apart in rows 42 inches (105 centimeters) apart all work out to be about the same amount of growing room per plant.

Where low soil moisture threatens to be a short-lasting problem, you might increase the spacing to nine inches (23 centimeters) apart in rows 48 inches (120 centimeters) apart. Using a row spacing greater than 48 inches makes little sense for most varieties (see the root system drawing). Where a severe shortage of rain during the growing season is a certainty and watering is not possible, the maximum amount of space you should ever give a single sweet corn plant is about 16 square feet (1.5 square meters) or four-foot centers, imitating traditional Native American gardening; growing in hills spaced four feet on center, putting four seeds in every hill, and then ending up with one plant (one for the worm, one for the crow, one to rot, and one to grow) by thinning progressively if necessary.

Because the ears are wind-pollinated and the pollen is heavy, when corn is grown in a single long row you may find many ears will be only partly filled. It is best to grow corn in a patch at least two rows wide, with at least six plants to the row. It is better to make the entire corn bed moderately fertile, rather than confining soil amendments to the rows or hills. However, extremely high levels of fertility aren't needed for this medium-demand vegetable.

Most varieties will sprout in slightly cooler soil than, say, beans will, but the plant is not frost-hardy. Unless you live in a short-season area, it is safer to wait until the earth is at 60°F (15.5°C) before sowing. If you're growing in rows, first spread soil amendments over the entire patch and dig them in, lay out the rows with a string line, and then poke a hole in the earth about two inches (five centimeters) deep with your finger and drop in two seeds. If two come up from every spot, thin to one plant when they are about four inches (ten centimeters) tall and securely established. If you get the odd blank spot, allow two plants to grow in an adjoining spot. Keep the patch well (and shallowly) hoed until the corn has shaded the ground, suppressing any newly emerged weeds with deep shade.

Pests and diseases. After the pollen drops, the silks become attractive to the corn earworm moth, who may lay her eggs there. This is the time to treat the silks with Bt if that pest is a problem in your area.

Varieties. Hybrids will outproduce the best-maintained classics by at least half as much again, mostly because most classic open-pollinated varieties have deteriorated to the point of becoming nonproductive. If you want to grow your own seed or believe it is better to use OP varieties, I suggest that at first you sow your main patch with hybrid seed and also trial a few classics. If one of these proves worthy of growing, you can make the switch your second year. The classic sweet varieties are more closely related to the old Native American field corns and usually are acceptable for making cornmeal, flour, or parched corn.

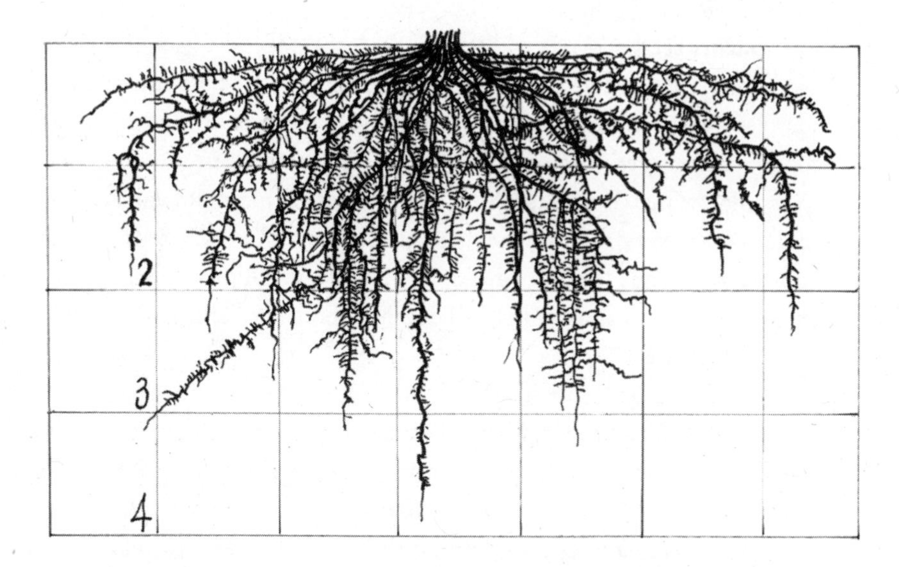

Figure 9.8: *A sweet corn root system at eight weeks old. (Figure 4 in RDVC)*

If your intention is to grow a large patch of field corn (not primarily for use as sweet corn) to use as a family staple, then I'd reverse my advice about hybrid varieties. Hybrid field corn appears to be more productive than the old classic varieties because it has hybrid vigor (this factor alone increases yield maybe 10 percent) and its careful, uniform breeding insures there will be no off-type nonproductive plants, but these uniform hybrid ears offer less nutrition to the consumer. Because it doesn't need to draw as many nutrients from the soil, hybrid field corn appears to grow better on less-fertile soil, but it will make more bulk containing less nutrition than the old-fashioned OP types. The classic OP varieties might fail on poor ground that the new hybrids can succeed upon, however, the hybrid's low-protein, low-mineral, high-calorie food is not something I'd want to maintain my family's health on. My advice is to do your own trials the first year; test as many sorts of OP field corn as you can find, determine the best one for you, and begin saving your own seeds the next year.

Harvest. Some varieties give a clear signal of ripeness: the wrapper changes color or the ear will lean out from the stalk. With others you have to peek into the end of the ear. The ear on the main stalk usually ripens a week ahead of those forming on tillers.

Saving seed. Corn is wind-pollinated; the pollen is heavy and won't blow one mile (1,600 meters) except in a gale so strong it would knock corn plants flat. Isolating varieties by a quarter mile (400 meters) will be good enough for home-garden seed if the second patch is not upwind in the direction of the prevailing winds. If that is the case, I'd suggest a minimum isolation of half a mile (800 meters). If you give your plants a bit more space than Everybody Else recommends (i.e., do what I suggest), they will probably tiller and their tillers will receive enough light to develop good ears. This will be particularly true of OP sweet corn.

When you are harvesting the main ear, it is possible to determine if that plant is productive and desirable or not. It is even possible to take a nibble out of the tip of each ideal-looking ear and see how it tastes, thus making flavor selections. You can allow the plants that show desirable characteristics for the gene pool to remain standing so that the secondary ears will mature as a seed crop. Break off the undesirable plants at the roots or knock them flat with your foot. Do not try to continue a sweet corn variety unless you can allow at least 50 desirable plants each year to contribute to the gene pool. To choose 50 good ones out of 100 or more requires a patch of at least 250 square feet (23 square meters). I have also produced quite acceptable OP sweet corn by saving seeds from Miracle hybrid.

Allow the seedmaking plants to stand until they have fully dried out, or, if the season is not long enough, harvest the ears shortly after the first frost, pull back the wrappers without tearing them off, and plait them, hanging the seed ears in a warm dry place in braids of 25 or so. This is the Native American method of saving seed corn.

Legumes — Beans and peas of all sorts

Everybody thinks legumes enrich soil by making nitrates. Actually, all nitrates formed in the roots are immediately incorporated into above-ground parts — leaves and then seeds. Thus legumes do not supply fertility to companion crops. However, if their tops are turned under while still green and lush, their decomposition does add significant quantities of nitrates for the following crop. Nitrates are made by specialized soil-dwelling microorganisms that beneficially colonize legume roots, forming nodules. These organisms won't be present in soil that is highly depleted of organic matter.

Growing details. Legumes need substantial levels of minerals, especially calcium (lime) and phosphorus, as well as having the nitrate-forming bacteria present in their soil, or they don't grow well. There's an old farmer's adage about this: Feed your phosphate to your clover, feed your clover to your corn (plow it in), and you can't go wrong.

When starting a new garden in humus-deficient soil, it is reasonable to assume that nitrate-forming bacteria are not present in sufficient numbers. In that case, inoculating legume seeds with these organisms might seem sensible. Farmers using depleted soil do this to avoid buying nitrate fertilizers for legume crops. Inoculants purchased in bulk are far cheaper than nitrate fertilizers, but buying inoculants in small packets can cost as much as or more than fertilizing a garden patch moderately. After the garden has received some manure or compost for a year or two, the needed organisms somehow appear. You can tell when this has happened by digging up a growing bean or pea plant and seeing if there are small pinkish-colored lumps on the roots; these are the nodules.

Consider all legumes low-demand crops, as described in Chapter 2; prepare their soil accordingly. Spacing recommendations for all types are found in Figure 6.1. Except for peas, it is a good idea to plant two seeds in every position you want a plant and then thin to the best plant after they are growing well. Peas, which are usually growing in a season without risk of severe moisture stress, need no thinning unless they come up severely overcrowded.

Peas are frost-hardy; some varieties are so remarkably hardy that they are used as overwintering cover crops. They can germinate in the chilly soils of spring, but it's best to chit the seeds if you're aiming for the earliest possible sowing. Bush varieties do not climb; they mature in concentrated fashion and are best grown in massed, multiple-rowed plantings across wide raised beds. Each spring I start three successive shelling pea patches, each containing five or six short rows, sowing a new patch every ten days. That way I get a continuous harvest lasting more than a month. Bush snow pea and bush snap pea varieties do not climb either. Because bush peas leave the soil in magnificent condition, I also sow large areas of them as a potentially edible green manure on any beds that can be worked in early spring but are not needed for a few months. If I need the bed before the peas are mature enough to harvest, I yank the vines and compost them. If green manure pea patches grow on to maturity, I somehow manage to eat, give away or freeze all those extra peas.

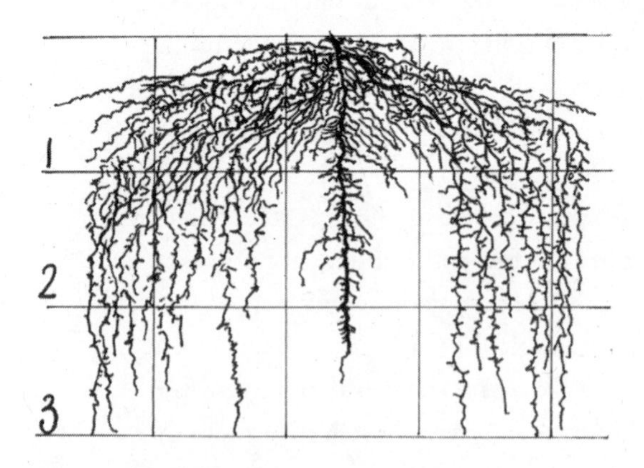

Figure 9.9: *A Tall Telephone climbing pea at maturity. Notice the incredibly dense and fine roots; is it any wonder they leave the soil in such friable condition? (Figure 52 in RDVC)*

Climbing varieties such as the Alderman or Tall Telephone, the original Sugar Snap, and a few rare but still surviving climbing snow pea varieties do allow more extended pickings and also have superior flavor compared to any bush pea. Climbing peas require a trellis at least six feet (180 centimeters) tall. I make my own each year by weaving bailing twine into a square-mesh fishnet of about nine inches (23 centimeters) on a side. First I stretch the vertical lines tightly between a top and bottom railing, and then I tie the horizontals to the vertical strings.

Fava beans (broad beans) are even more frost-hardy than garden peas. In mild climates they are sown in autumn and harvested in spring; where the soil freezes solid they are sown early in spring, like peas, but they take a bit longer than peas to mature. In maritime or mild climates it is wise to sow them as late in autumn as the seeds will germinate in order to keep their stalks as short as possible over winter because when beans finally begin to set in spring, the pods may form so high up that the top-heavy stalks will fall over.

Fava beans usually become diseased in hot weather. Large-seeded varieties are mainly used as shelling beans; the small-seeded types make the most excellent cool-season green manure and sometimes are grown for seeds to be used as animal food.

Snap beans (French beans) are not frost tolerant. In climates with cold winters, sow them in spring once the soil has reached 60°F (16°C). Where summers are hot and steamy and winter almost nonexistent, sow again after the worst heat of summer has passed for harvest in autumn. Bush varieties yield for only a month at most; climbing varieties take a week or two longer to begin yielding, but they usually continue bearing as long as their roots have room to continue growing and the weather suits. Bush varieties have much smaller root systems than climbers and consequently are far less drought tolerant. Climbing varieties also have superior flavor; I prefer eating them. In my garden I put in a small planting of a quick-maturing bush variety to fill the gap before the climbing beans start bearing.

Climbing beans are more work to grow than bush varieties, but easier to harvest by hand. They may be supported by tripods made of eight-foot-tall (245-centimeter) skinny rough poles with bark and twigs still attached, lashed together

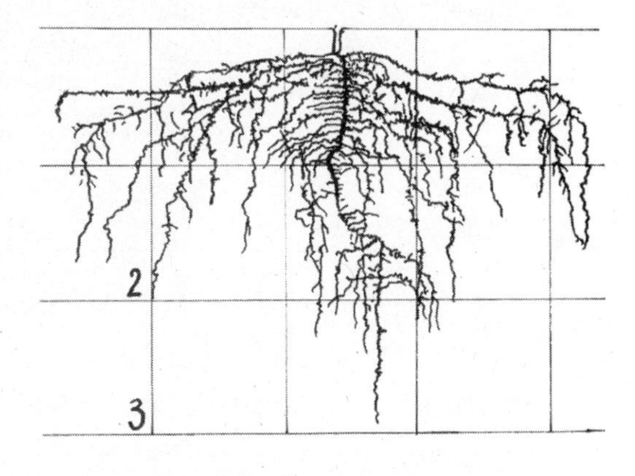

Figure 9.10: *A bush snap bean when the flowers are beginning to set pods. (Figure 55 in RDVC)*

near the top as though for a tepee. One plant grows up each pole. In the Willamette Valley, where for decades nearly all of America's canned green beans were produced, Blue Lake pole was the variety of choice before labor costs made it necessary to grow the less-tasty machine-harvested bush types. To support the vines, stout and well-braced seven-foot-tall (210-centimeter) posts were erected at the ends of each row. A strong tight wire was stretched between them from top to top, and another wire was stretched five or six inches (13 to 15 centimeters) above the ground. Props were inserted as necessary to hold the massive weight that developed on a top wire that was hundreds of feet long. String was then run top to bottom to top to bottom,

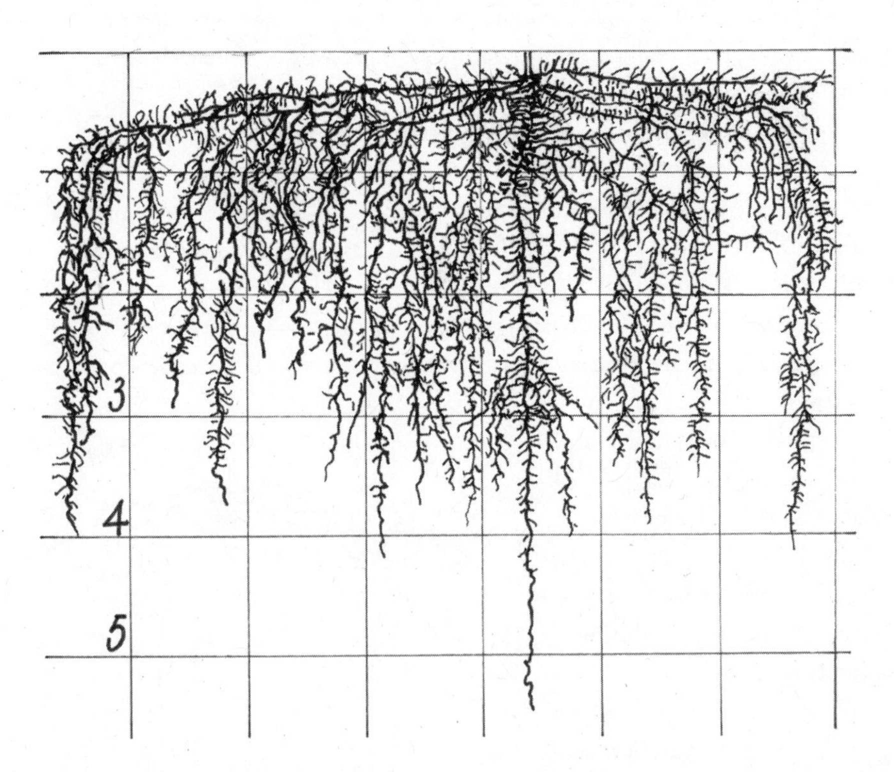

Figure 9.11: *This is a climbing bean whose pods were not picked but were grown for mature seed. The drawing shows the root system at full development; new pods have just stopped setting and the seed load is beginning to mature. Clearly, if given growing room, climbing beans are better adapted to dealing with moisture stress than are most bush varieties. (Figure 58 in RDVC)*

etc., in "VVVV" shapes. The bottom of each V was about 12 inches (30 centimeters) apart, with one plant coming up at the base of each V. High-school kids used to be happy to get jobs stringing beans. The plant spacing indicated in Column 3 of Figure 6.1 is based on this method. There are many other support variations possible. The most important thing about climbing bean structures is to allow the plants enough growing room; the vines can fork many times, and one plant may produce quite a few runners.

Picking snap bean pods before seeds start forming in them increases the number of new pods set.

Runner beans (Phaseolus multifloris) grow more aggressively than climbing snap beans. I once entirely filled a seven-foot-tall (210-centimeter) boxy structure covering an area of four by eight feet (120 by 240 centimeters) with the vines resulting from sowing only three runner bean seeds. Scarlet Runner, the traditional Native American variety, was mainly grown for shelling beans and is not a great variety for producing edible pods, but the British refined this species into something that produces the most elegant-looking snap beans with the richest flavor. I recommend one called Prizewinner; it is aptly named. *P. multifloris* has difficulty with hot weather and does best in maritime climates where summer nights are cool. An effective way to make a far higher percentage of flowers set pods is to regularly spray water on the profusely blooming vines shortly before dark.

Dry beans and shelling beans are the same species *(P. vulgaris)* grown as snap beans. Anyone who cooks even a few pounds of beans a year might grow their own, if only because fresh beans (less than one year old) cook a lot faster and taste much better; what is sold in the supermarket is often of indeterminate age. You can cook the seeds of any snap bean; many are delicious, though some are only "acceptable." Varieties bred for the purpose are usually best. Almost all dry bean production these days is done with bush varieties. These are grown exactly as though growing snap beans, but the beans should not be harvested until the crop has dried in the field. The old Kentucky Wonder Brown Seeded (climbing type), sometimes called Old Homestead, also has delicious seeds and, given sufficient growing room, is more drought tolerant than most bush types.

Ideally, delay harvest of dry cooking beans until the odd mature seed is shattering out of its pod. But harvest in a rush if nearly ripe seeds are about to be rained on. Pull the plants, roots and all, shake off the soil, and then spread them out under cover with good air circulation to finish drying down. When they are crisp, you can thresh them in various ways: grip a plant by the lower stalk and vigorously bang it back and forth inside a metal oil drum; spread the plants on a tarp and hold a dance; patiently shell the dry pods one by one. To clean the chaff, winnow the seeds between two buckets in a stiff breeze. Make sure the seed is fully dry before sealing it up. It's best to store large seeds in something that will pass moisture and air, like a paper sack or old feedbags (they respire quite a bit of oxygen).

Vegetable soybeans have seeds that are eaten green, like shelling beans. Soybean varieties are quite specific to a narrow range of latitudes because they grow vegetatively until the daylength orders flowering and seed formation. At this point the plant completely goes into bloom and makes one set of seed that ripens all at once. To get an extended soybean harvest, you must grow several varieties of staggered maturity. You can sow all varieties at the same time. Obviously, all other things being equal, the later seed formation happens, the larger the bush will have grown, the more flowers it may form, and the larger the harvest will be. But if you use a variety bred to grow closer to the equator than you are, it might start seed formation too late to mature seeds.

To prepare vegetable soybeans for the table, wait until most of the seeds have become fat but are still soft and tender. Cut the whole plant just above the soil line, stuff it into a huge pot with an inch (2.5 centimeters) of water on the bottom, and steam it for about five minutes. It's cooked as soon as the seeds pop out of the pods when the pod is gently squeezed. Don't overcook! Remove the whole thing and put it into a big bowl on the center of the table for the whole family to share, like hot popcorn. Tug the pods off the stem and then pull each pod through your front teeth, popping the seeds into your mouth. Traditionally, the whole plant is lightly salted when it is taken from the steamer, like popcorn.

Dr. Alan Kapuler of Peace Seeds in Corvallis, Oregon, does much work with edible soybeans; so does Johnny's Selected Seeds.

Lima beans grow excellently in heat but, surprisingly, do not set seed well in really hot weather. Where summers are long, hot, and steamy, sow lima beans in spring for harvest before it gets really hot, and then sow them again about two months before the weather cools down, for autumn harvesting.

Louisiana State University agricultural extension service says small-seeded lima varieties, locally called butter beans, do better in the Deep South of the United States. LSU also strongly recommends that home gardeners grow climbers because they may be picked for a longer period.

Southern peas produce only where nights are warm. Each sowing matures a batch of shelling peas more or less simultaneously. If you want a continuous harvest, you'll have to sow a new patch every few weeks. There are two general sorts, vining and bush. The vining type needs a wider spacing and must be hand-harvested. The bush types were bred for commercial production and do not have the flavor of the old standards. If planted much more densely than for eating, southern peas make an excellent and quick-growing summertime green manure.

Pests and diseases. There is an insect called the bean weevil (the pea weevil is different, but a similar species) whose larvae tunnel into ripening seeds, pupate inside, and then tunnel out, emerging as little crawling insects. Each seed that hosted one will have a small exit hole in it. The seed usually germinates successfully, but it has lost a bit of its food supply, consumed by the larvae and emerging adult. These insects are annoying but not catastrophic and may be stopped in their tracks before emergence by freezing the seed for a few days shortly after harvest. I have cooked beans that grew weevils to adulthood and they tasted okay. These critters do not exist Down Under, which is why bean and pea seeds may not be freely imported.

Saving seed, all types. Many legume species rarely cross, requiring no isolation. However, favas are often bee pollinated; in an organic garden, peas, runner beans and garbanzos are often pollinated by other insects, so if saving seed on these, it is wise to grow only one variety in your garden.

For the purposes of growing just enough seed to plant a small garden and also to keep a legume variety pure, it's best to save seed from only one or two perfect plants. If you'll be needing more than an ounce (30 grams) or so of seed for any variety, allow two seasons to increase your seed stock. The first year, start by saving the seed from only one ideal plant; the second year, grow that seed separately in what breeders term a "line," and keep a close eye on that line. Remove any plant that seems to be off-type (this is called "roguing" in the seed trade). The seed you harvest from the perfect plants in that row or line (and with luck all your plants will be perfect) is your planting stock for the forthcoming few years. High-quality breeding would reject any line that was not absolutely uniform.

When I grow bush peas of any sort, I save seed from the inevitable overlooked pods. When these have three-quarters dried out on the vine, I pick them and bring them into the house to completely dry out. Doing this gives me twice as much seed as I need for the next year which is why I can afford to use garden peas as a spring green manure crop. I never overlook pods when I harvest climbing peas, so instead I allocate two or three of the vertical strings on my netting for seed production and pick the rest for the table. I do the same with climbing beans.

Okra

It's been many years since I've lived in a place hot enough to grow this vegetable; I sorely miss it.

Growing details. Okra needs heat and prefers humid conditions. The plants can get quite large and will continue bearing as long as they can continue growing. The seeds are difficult to germinate, so it is wise to sow four seeds about three quarters of an inch (two centimeters) deep at every point you will want a plant and then progressively reduce competition to a single plant. Do not sow until the soil has warmed up to at least 70°F (21°C), and it's best to soak the seeds overnight before sowing.

Okra does best if it is not put in super-fertile soil; too much nitrogen makes it set far fewer pods. However, like all plants that will fruit continuously, once it has been bearing for a while it does better if it is given a bit of fertigation or a light side-dressing of COF. Okra makes a deeply penetrating taproot and an aggressive system of surface feeder roots; it can survive drought, but to keep bearing heavily it needs more than minimum moisture. In dry spells this is a perfect crop for periodic fertigation.

Pests and diseases. Grow the crop on a spot that hasn't seen it for many a year. Okra is sensitive to attack by soil-dwelling nematodes. These tend to build up in soils that grew sweet potato or squash, so do not follow these crops with okra the next year. The best garden crop to precede okra is corn, which lowers nematode levels. Leaf-eating insects may bother the crop, especially damaging small young plants when the weather hasn't warmed up enough to suit. Once rapidly growing plants have attained some size, they will usually shrug off a little nibbling. When they are small, spraying rotenone/pyrethrum or Bt (for loopers) might get them past the danger point.

Harvest, storage, and use. Initially the pods form rapidly. While enlarging fast, they are tender. After the initial growth spurt slows, they become woody. Size does not indicate the state of the pod, because if the plant is growing lustily, the initial rapid growth spurt might create pods over five inches (13 centimeters) long. Do pick pods before they start getting crunchy. It's best to do that by cutting them off with a sharp knife. Some people are sensitive to the plant's tiny spines and should wear gloves when picking.

It seems to me that okra is inevitably overcooked. I suggest you try eating a raw pod in the garden the instant it is picked. In my okra patches, hardly a pod ever made it past the harvester and into the kitchen. If chilled below 45°F (7°C), the pods rapidly deteriorate; if not cooled, they rapidly become woody, so it's best to eat them promptly. Harvesting must be done almost daily.

Saving seed. If you have a large patch — over 25 plants — why not save seeds? Okra is bee-pollinated, so to preserve a pure variety requires isolation of over half a mile (800 meters), though pure varieties are not essential for producing lots of edible pods. On every plant that has mostly desirable characteristics, allow one unharvested pod to mature and dry out. The seed develops rapidly. Do not let more than one pod per plant form seed unless you are only growing a seed crop; unharvested pods greatly reduce formation of new pods.

Chicories — Endive, escarole, radicchio

Chicories are remarkably cold-hardy salad greens, but most can't withstand frozen soil. In cold-winter climates, chicories are harvested in autumn. In hot climates with mild winters they are a cool-season crop. In maritime climates they are good eating in both autumn and winter, although there can be disease problems caused by winter humidity. Rarely are problems caused by frost.

If you believe endive and escarole are horribly bitter, you probably tried some that had not experienced any serious chilling. Frost kills plants because when their cells freeze, they burst. Some plants survive frost by converting starch in their cells into sugar. A solution of sugary water freezes at a much lower temperature than plain water does; the sugar acts like antifreeze in an automobile's radiator. After the temperature comes back up, the sugar remains and the flavor is much better. This same thing happens to kale, Brussels sprouts, and many other leafy crops when they get frosted but not killed.

Growing details. Chicories may be considered low-demand vegetables that will grow faster and bigger when given more than the minimum. But it doesn't seem to matter how large or how small they are when they arrive at the cool season; their eating quality is about the same. I suggest that if you're growing them more slowly on poor soil, sow them a bit earlier. Seed is usually quite long-lasting; if you don't get effective germination, you were sold some really old stuff. Thin chicories progressively; try to place the seed so their row starts out with about one seedling per inch (2.5 centimeters).

Chicory root systems resemble those of other biennials like beetroot or carrot, penetrating subsoil in search of moisture and nutrition. They are capable of surviving drought and coming out the other side of it in the cool season with some salad greens for you.

Pests and diseases. In maritime climates, chicories tend to develop mold in winter. The heads may rot back to stumps. In really wet winters they will die; in milder, dryer years they grow steadily right through winter and into spring, when they go to seed. Erecting an open-sided roof of clear plastic over the plants almost guarantees winter survival and much higher leaf production. Talk about frost-hardy! I have seen endive and escarole shrug off nighttime lows of under 10°F (–12°C) — but the soil was not frozen).

Varieties. Lately, productive and uniformly heading radicchio (with small, round, red, cabbage-like heads) have appeared. Radicchio usually heads in autumn in cold-winter climates and probably would be a good cool-season crop in mild climates. Endive and escarole are basically non-heading radicchios, but one has broad thin leaves and the other highly frilled stemmy leaves. There are other *haut cuisine* chicories, too. One, in English called Sugarhat and in German, Zuccerhat, makes a conical and firm head a bit like a Chinese cabbage. I find it rather bitter. Also, a few hybrid heading chicories similar to radicchio have begun to appear recently. Most chicory varieties are open-pollinated and so far I see no need to use expensive hybrid varieties.

Harvest. In climates where the soil doesn't freeze, permit a few outer leaves to remain alive on the stem when you are cutting the heads. If you do this, the plant will not die but will start producing a whole new batch of smaller leaves that may be picked later in winter. These will be more tender than the big heads. In climates where the soil does freeze, you can dig out endive and escarole a week or two before the soil starts freezing. Remove their largest outer leaves, shake the soil from their carrot-like root, and then transplant them into tubs or beds in the root cellar, to be used during winter. If these plants survive winter in the cellar, you may replant them outside in spring for making seed. If only a seed crop is intended, you can trim the tops (without damaging the crown and its growing point) and cellar only the roots overwinter, stored like carrots in moist sand.

Saving seeds. Chicories are biennials. After overwintering, the plants put up large masses of waist-high jointed stalks covered with pretty blue flowers that almost always self-pollinate. However, it's a good idea to isolate varieties by 50 feet (15 meters). Chicories have a tendency to mutate, so it's best to save seed from a single good plant, much as I suggested for saving bean seeds. The capsules holding chicory seed are quite hard and difficult to shatter. If you hold a vigorous tap dance on thoroughly dry stalks spread on a strong tarp on a hard surface, a goodly portion of the seeds will be broken loose. After sieving out the larger trash, you can winnow the seeds from the fine chaff. After throwing chicory trash into the compost heap, I've had zillions of endive and escarole plants coming up all over the garden — a not undesirable bonus.

Crops that are harder to grow

Lettuce

To achieve the best eating quality, lettuce must grow fast and unstressed. Its root stores water, but when lettuce has to survive by drawing from its moisture reserve, new growth stops and the leaves become bitter and tough. Its soil has to be fertile and the plants must not be crowded; competition slows growth. Any leafy plant needs nitrates to grow. To grow as many leaves as rapidly as lettuce should demands high levels of available nitrates. So when preparing its growing bed, consider lettuce a high-demand crop.

Most varieties don't tolerate much heat; lettuce seeds won't even germinate in hot soil, much less grow. Thus, where summer is hot, consider lettuce a quick-growing cool-season crop. Fortunately it is frost-hardy; some varieties can harden off to become remarkably frost tolerant, especially when half-grown, but I know of no variety that will withstand short-term freezing the way kale can.

Growing details. The most important thing, besides giving lettuce the soil conditions it requires, is to thin progressively, carefully, and promptly. Competition slows growth; don't let it happen.

Varieties. Dark green looseleaf varieties grown on properly balanced mineral-rich soils have been tested at nearly 20 percent protein, almost as high as some forms of flesh food. And unlike meat, salads are eaten raw, so are far more digestible than cooked flesh. Iceberg and other ball-heading types with blanched cores, lacking high-protein chlorophyll, are much less healthful.

Harvest. Growing salad mesculin — a densely spaced salad greens mixture that is cut repeatedly, like mowing a lawn — has become a popular practice. The catalog suggests that from a single sowing you can harvest for months. But I believe that after the first cutting, mesculin usually results in second-rate salads. Of course, if you spread a strongly seasoned fatty dressing on your greens, you won't notice how bitter and dry they have become.

I prefer to grow looseleaf heads to full size so that their small inner leaves semi-blanch, as they're supposed to, and then cut the entire head at the stem. I don't allow mature heads to sit around uncut, wasting valuable space in my garden, because when they are only a few days too old they rapidly become bitter. I compost old lettuce. If you have chickens or rabbits, I'd suggest giving it to them — they like it. A single sowing of one variety remains in prime eating condition for only seven days in warm weather, so I plant each lettuce patch with several varieties of differing maturities, sown at the same time. And I make repeated sowings, usually about three weeks apart. Three weeks is about the longest period I can stretch the harvest of assorted varieties made on a single sowing date.

We try to eat one huge leafy green salad every day that it is possible to produce a green salad. In my climate, that's 365 days a year. Our salad is large enough to be a meal in itself. Usually half of that salad is lettuce, so the kitchen needs about one mature looseleaf head every single day. One looseleaf head occupies about ten inches (25 centimeters) of row. Three weeks' worth of lettuce, 21 heads, is about 18 row feet (5.5 meters). The short rows across my raised beds are about four feet (120 centimeters) long. So spring through summer, I start five such rows every three weeks. And going into the cool season, when the entire garden grows much more slowly, I'll make larger plantings because when the sun is weak, nearly mature plants will slowly size up for weeks.

Saving seed. Lettuce is a self-pollinated annual that grows vegetatively for a time and then has seedstalks emerge (when the stalks emerge, the lettuce is said to be "bolting"). These become covered with tiny yellow flowers; each flower capsule forms only a few seeds. After the seed has ripened, the plant dies. Success at making seed depends on how warm and how long summer is where you live; it may also depend on how cold winter is. My winters are mild and occasionally frosty. So I sow my seed crop in mid-autumn and overwinter the seedlings outdoors without protection. In harsher locations the seed plants can be started under glass at the first signs of spring and set out the earliest possible date. Either way, the heads must be fully developed and probably must be bolting before summer heat arrives if the plants are to mature their seed load. Most varieties reach a critical point about six to eight weeks after the summer solstice when, prompted by shortening days, plants forming seed will switch over from ripening that seed load to growing vegetatively again. If your seed crop hasn't matured by then, it won't.

Although lettuce flowers are almost always self-pollinated, there is a slight tendency for crossing. If perfect purity is a big concern, isolate varieties by 50 feet (15 meters). If you are willing to yank out (or eat) anything visibly off-type, then don't worry about isolating varieties and don't be concerned if one plant in a thousand is a cross. Who knows? That unexpected cross might be bred into a new variety you'll treasure.

The seeds ripen irregularly and tend to shatter, so as soon as you can see mature seeds on a particular branch, cut off that branch and place it on a tarp in an outbuilding to mature. When you have picked all the seedstalks and they are pretty much dried, drag the tarp into the sun and let the stalks dry to a crisp, then rub each stalk between your (gloved) hands, breaking up the dry flower capsules. Discard the twigs and straw. Put the dusty remainder into a large bucket and (wearing gloves) rub it between your hands until all flower parts have been fully powdered. Then winnow out the seed on a day with a gentle and steady breeze. There is not much difference in weight between the dust and the seed; if the breeze is too strong, you'll lose most of your seed. My yields are not high. I suspect if I were in coastal southern California, where nights are warmer (and where most commercial lettuce seed is grown), I'd get double or more the yield I get on Tasmania. I try to let at least two plants from each variety I grow make seed every year. I give lots of it away as gifts.

Arugula

Arugula is a near-wild leafy annual in the cabbage family. In small amounts it perks up salads. When cut early, before the flavor gets too peppery, it can be made into a salad all by itself. Buy a large seed packet because it must be sown frequently. The seed should be inexpensive.

In hot weather the leaves go from being pleasantly mild to being too strong to eat in under two weeks. If you want to harvest much before it turns bitter, give the plants fertile soil. I start about four row feet (120 centimeters) of arugula at least every three weeks. When I sow my lettuce patches, one extra short row is always arugula. Sprinkle the seeds thinly into the furrow so the seedlings touch at emergence, then thin progressively as you wish to eat it. Small seedlings are especially delicious.

In hot climates it is best to grow arugula in spring and again after summer's heat breaks. I overwinter my seed crop, but where winter is too harsh for that, spring sowings will mature seed in summer.

Parsley

I'm saddened when I see a corn-and-tomato gardener buying parsley seedlings in a garden center. This vegetable is not hard to grow from seed if only you know how to make the seeds come up. Pregermination helps, but is not essential. What *is* essential is starting the seeds earlier than someone who "puts in their garden" over one weekend in late spring can do. Midspring, about the time the apple trees are blooming or the daffodils have nearly finished, sprinkle the seed about a quarter inch (six millimeters) apart in a furrow and, if possible, cover it with a bit of finely textured mellow compost instead of soil. Germination is slow; it can take as long as three weeks, but at that time of year the soil naturally stays moist. If it doesn't, you'll have to keep it moist; water the row. Progressively thin

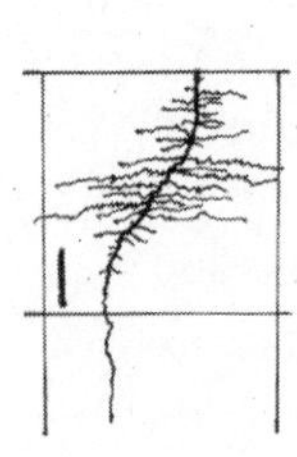

Figure 9.12: *Leaf parsley at 60 days after emergence. Note how weak and slow-growing the root system is. Parsley is the only relative of the carrot that needs a bit of coddling. (Figure 65 in RDVC)*

until the plants stand about four inches (ten centimeters) apart. Then pluck leaves as wanted.

Parsley does not aggressively mine the subsoil as its relatives carrot and parsnip do. If you're not getting as much production as you'd like, side-dress the row with something containing a good deal of nitrogen and make sure the plants are getting enough moisture. In mild-winter areas, parsley may continue producing all winter. In hot-summer areas it may have to be started after the heat of summer has passed. In that case, erect temporary shade over the rows and keep them moist.

Parsley is a biennial and a relative of the carrot. Make seed as you would for carrots.

To grow the best parsley leaf, think of it as a fragile root crop that makes edible leaves. There is a variety often called Hamburg that is grown for its edible roots, not the greens. These taste much like parsnip and may be eaten raw, grated into salads, or cooked. Grow these slow-to-mature roots as though producing carrots that need more growing space, maybe 5 by 18 inches (13 by 45 centimeters).

Carrots

To my thinking, the carrot typifies the survival strategy of all biennial root crops — sneak a deep tap root through the opposition; then begin drawing on moisture and nutrition located deeper than its neighbors can reach. A supply of raw materials assured, it fills a storage chamber with surplus food. Then it rests over the winter. In spring it starts with a huge nutritional storehouse that it uses to rapidly overtop its neighbors and shade them out, then begins making seeds.

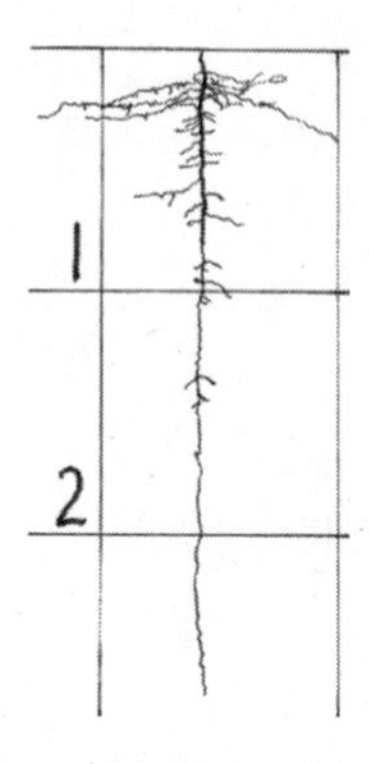

Figure 9.13: *A chantenay carrot seven weeks after emergence. (Figure 61 in RDVC)*

Growing details. It is easy to grow carrot tops; the edible part is not quite so easy. If you put in the necessary work at the right time and use decent seed, then eight out of ten plants should make a near-perfect carrot. If that isn't your result using quality seed, then it is probably your fault. To succeed, keep this principle in mind: To be sweet and tender the root needs to develop rapidly; slow-growing, struggling plants make woody carrots that often are bitter. Three things contribute to rapid growth: enough soil fertility (but not too much or you'll end up with more top than bottom); open, loose soil, especially a top foot (30 centimeters) that graciously allows the root to swell up as fast as it wishes to; a steady moisture supply in the subsoil. It's best if the plant doesn't have to struggle against too much competition for this third requirement.

Commercial crops are grown on light, sandy ground that is naturally open and loose. In hard soil, carrots must develop more fiber so as to be able to force the soil aside, making them tougher and irregularly shaped. If your soil tends to become compacted, you'll need to dig in a large quantity of compost or well-rotted manure, perhaps as much as a two-inch-thick (five-centimeter) layer. But beware of using potent compost; if the layer of soil that the carrots themselves form in is too fertile, they can become hairy and may fork. Your carrots need to discover most of their nutrients in the subsoil.

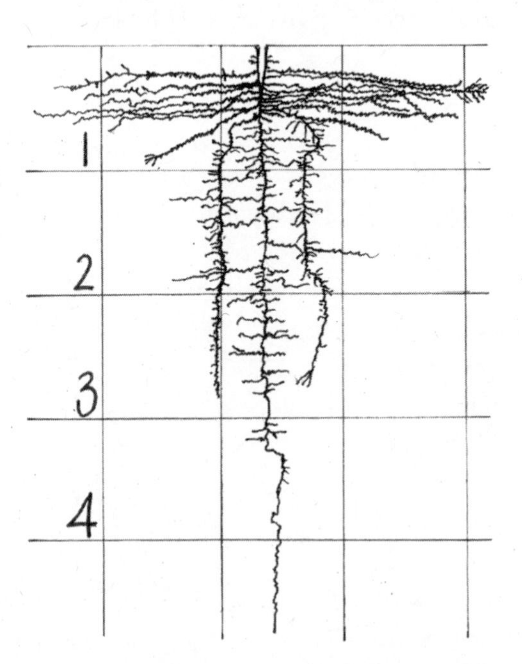

Figure 9.14: *The same carrot at 11 weeks, at a size most would consider barely large enough to eat. (Figure 62 in RDVC)*

Providing the crop with a steady supply of moisture has more to do with thinning than with watering. Thinning *is* success with carrots. First, try not to sow too much seed. I thoroughly blend a heaping half teaspoon (2.5 milliliters) of strong carrot seed into a quart (liter) of fine sand or compost, then uniformly sprinkle this mixture into the bottom of 25 feet (8 meters) of furrow. I end up with evenly spaced seedlings standing about half an inch (1.25 centimeters) apart. If yours come up thicker than this, do some preliminary thinning right after germination to get them closer to this density. Don't overthin, though; many seedlings die in their first month. As they grow, thin the row progressively, steadily reducing competition. About the time the tops are three to four inches (seven to ten centimeters) tall, there should be no more losses due to diseases. Then you should thin the row so that when the carrots mature, they will just touch. How far apart this will be depends on the size of your variety. Some are the size of your fingers; others may be as much as three inches across the top at maturity.

Varieties. Tender, juicy varieties bred for munching don't have enough fiber to store well or grow large without cracking. The classics of this sort were called Nantes types. Now they have many different proprietary names. If you are growing carrots on heavier soils that require they have sufficient mechanical strength to push aside resistant earth when the roots are filling out, grow more fibrous types — traditionally called danvers, chantenay, berlicum, or flakee, but these days often given proprietary names. These are also better for overwintering in the earth or for storage in the cellar.

Harvest and storage. If you do not have a younger carrot patch coming along, harvest by removing every other one in the row. This is not always possible if the variety you're growing has weakly attached tops; these sorts may only be harvested by digging. You will be amazed at how large some varieties of carrot can become while still remaining tender and sweet when they have little root-zone competition.

In mild climates, carrots will overwinter in the earth. In places where it can get severely frosty, covering the crowns by hilling up a few inches of soil over them will serve as protection. Where the soil freezes, the roots may be root-cellared in a barrel of moist sand.

Saving seed. The wild carrot is commonly called Queen Anne's lace. The garden carrot was domesticated only recently, less than 500 years ago in fact, and crosses freely with Queen Anne's lace. If there is none growing in your vicinity, you may grow carrot seed.

In spring, replant overwintered roots, carefully hand-selected for perfect configuration, about one foot (30 centimeters) apart in rows at least two feet (60 centimeters) apart. They will shoot and make flowers that will eventually turn yellow-brown as they curl up, wrapping the mature seed inside. Cut the flowers one by one as they are nearly dried out (before any seed is lost) and put them in a large open bucket or toss them onto a tarp in the shade under cover. When fully dry, put them in the sun one afternoon to dry to a crisp and then rub the flowers between (gloved) hands to separate out the seeds. Winnow out the dust and chaff. Carrot seed that did not get rained on while drying down and that fully matured before being cut will last many years if

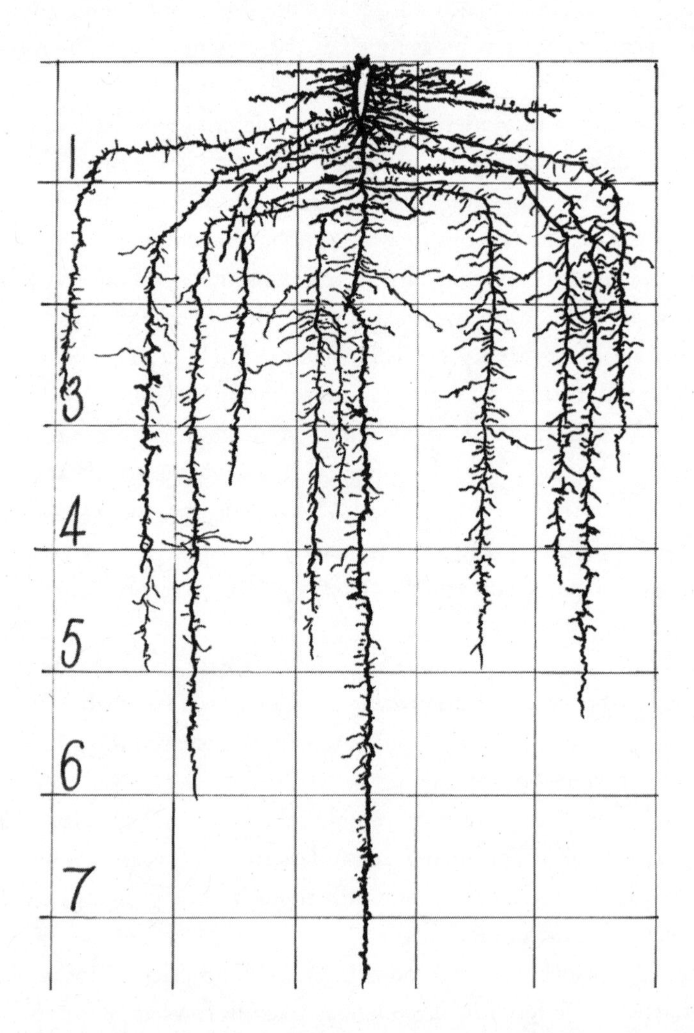

Figure 9.15: *The same carrot at 15 weeks. If the carrot were not competing with other carrots, it might weigh nearly half a pound (220 grams) at this point; because it never stopped growing rapidly, it still would be sweet and tender. If allowed to continue growing, that carrot might weigh a pound (450 grams) by the time frost came and would still be good eating. (Figure 63 in RDVC).*

given decent storage. Thus it is possible for one gardener to grow seed for several varieties. Because the species is bee-pollinated and depends on outcrossing to maintain its vigor, do not make seed for more than one generation unless you are able to include at least 50 plants in the seedmaking process. A 25-foot-long by 4-foot-wide (762 by 120 centimeters) raised bed with a double row of carrots making seed will produce enough seed to last the neighborhood for a decade.

Parsnips

Parsnips are like carrots with differences. Most varieties are longer than carrots, which means they need a deeper layer of loose topsoil to succeed. They have a slightly lower need for fertility and grow more slowly than carrots, but do not crack when quite big as some varieties of carrot will. The biggest difficulty is getting short-lived parsnip seeds to germinate. Most seed lots are faltering only two years after harvest. Growing your own parsnip seed is a good idea. I sell my extra seed at the local garden center. My neighbors buy it eagerly because they know it will come up! It also grows a uniform crop because it is produced from hand-selected roots. Each year this sale earns enough to pay for all that I wish to buy at the garden center.

Growing details. Fresh strong parsnip seeds sprout slowly; rarely more than 25 percent emerge, so get quality seed in a big packet and sow it thickly. Cheap stuff makes mostly misshapen roots if it comes up at all.

If parsnips are started in spring, when naturally moist earth would make it easy, they'll have become so enormous by autumn that they're almost inedible and may be showing damage from insects and soil diseases. I start them on or shortly after the solstice so they size up when weather gets cooler. Gardeners living closer to the equator might start their parsnips closer to the equinox and consider them a crop that grows during the cool season. It can be daunting to get the seed to sprout in the heat, and the flavor of parsnips improves a lot after they go through some stiff frosts. Sow thickly. It helps to shade the furrows or to cover the seed with compost instead of soil. Pregermination and fluid drilling also help. Some gardeners place thin wood planks over the furrow and check under them every day; as soon as some seedlings are emerging, they remove the planks. Watering every day really helps and, again, sowing thickly helps. Then thin progressively and otherwise grow them as you would carrots.

Harvest and storage. Parsnips usually must be dug. When the earth gets close to freezing, the whole patch may be dug and the roots cellared in a barrel of moist sand, as carrots are. Some gardeners leave a part of the crop in the earth over the winter and find that when the soil thaws in spring they can still be dug and are quite good eating.

Saving seed. Save seed as you would for carrots. To avoid inbreeding depression of vigor, use at least 25 roots in the seed crop. Grow seed only from carefully hand-selected roots that show exactly the traits you want. Segregate the strongest and largest seed from the main central flowers and use that for yourself; give the rest away (or sell it). The seed falls from the flowers when it's fully dry, much as dill seed does, so cut off individual flowers one by one at the point when one or two seeds have detached. Let the flowers finish drying in a big bucket or spread them on a tarp under cover.

Radish, garden kohlrabi, turnip, and rutabaga (swede)

These vegetables are so similar that their cultural differences can be described in a few lines.

Growing details. All are medium-demand crops. All appear to be root crops, but kohlrabi is actually a swollen stem forming a ball-shaped vegetable above ground. Accurate spacing and a steady supply of soil moisture are essential for all of these crops. Except when using costly hybrid kohlrabi seed, I sow in furrows, then thin progressively and accurately to enough separation that the edible parts will not quite bump at maturity. Turnip and rutabaga seeds are tiny, necessitating shallow sowing, but they usually sprout rapidly and vigorously. If you are sowing in hot weather, it is best to cover them with fine compost and water daily.

Salad radishes are annuals; shortly after bulbing they go to seed. In hot weather, radishes go to seed even more rapidly, taste more pungent, and quickly get pithy when oversized, so in most cases the salad radish is either grown as a spring crop or is sown late in summer for harvest in autumn. In mild-winter regions it is sown in autumn for winter harvest. Make sure that by the time radish tops are about three inches (7 centimeters) tall, they are at their final spacing. If you treasure radishes, I suggest making two sowings about 10 to 14 days apart in spring and two or three sowings a few weeks apart at the end of the season.

Winter radishes act like biennials; they go to seed after overwintering. Do not sow them until several weeks (or more) after the solstice or they may go to seed in autumn. They form large roots that have thick protective skins. My favorite is Black Spanish. Peeled, coarsely grated, and dressed with a bit of olive oil, finely minced onion, and black pepper, it makes a great salad. Black Spanish needs a spacing of about 6 by 18 inches (15 by 45 centimeters); most winter radish varieties need this much room. In regions with freezing soil, most types are quite cellarable.

Turnips, like radishes, are spring and autumn vegetables; they're cool-season crops in hot places with mild winters. But if you grow them for spring harvest, you'll have to manage their tendency to go to seed before sizing up by encouraging them to grow extremely rapidly (and by harvesting them) before they bolt. When I was new to this game, I grew spring turnips for the challenge of it. Now I don't struggle and only sow them at the end of summer. Turnips will root-cellar; in my mild-winter climate they withstand the autumn/winter frosts while sizing up. But if left in their beds too long into winter, they become pithy and dry. The Asian hybrid varieties are excellent eaten raw.

Rutabagas (also called swedes) are an autumn crop that in mild-winter climates have an excellent ability to stand in the garden through the entire winter, remaining in good eating condition until they make seed in spring. They'll get quite a bit larger than turnips; I thin them to about eight inches (20 centimeters) apart in the row. Rutabagas are probably the most cellarable of the lot.

I lived in British Columbia for a few years and discovered rutabagas in Canadian markets were thinly coated with wax, such as is often put on the skins of cold-storage apples. I suspect the home gardener with a root cellar could melt paraffin or beeswax floating atop a large kettle of hot water, thin it by blending in a bit of vegetable oil, let the water cool somewhat (but not to the point that the paraffin became solid again), and then quickly plunge roots in and out, sealing their pores to prevent them from dehydrating. Rutabagas make a welcome change from spuds.

Kohlrabi, the small-sized variety, is a difficult spring crop — not because it goes to seed like turnips do, but because it rapidly gets too woody to eat in hot weather. But as an autumn crop there is nothing better! Kohlrabi is another vegetable you should not grow from cheap seed, of which there is too much about. If you start with a top-quality (usually hybrid) variety and then grow it right, 98 percent of your plants should make near-perfect balls the size of grapefruit, sweet and crisp, tasting better than the giant varieties. I sow hybrid kohlrabi in clusters of three or four seeds, half an inch (1.25 centimeters) deep, on stations of 5 by 18 inches (13 by 45 centimeters). Then I thin the clusters progressively so that by the time they have three true leaves, only one plant stands at each spot. I first sow about 30 days before the weather will begin to significantly cool down; thus by the time the bulbs begin forming, the weather suits them. With proper genetics, the bulb will slowly enlarge in cool weather without getting woody, slowly enlarging. Beware: Some commercial varieties are bred for a quick once-over harvest and don't hold long after reaching marketable size before becoming woody.

In my mild-winter climate, I make a second sowing to size up after the weather turns really chilly. This lot remains in prime eating condition until the spring warm-up begins. Kohlrabi is also cellarable if the roots are left attached, the large leaves are cut off to facilitate air circulation, and they are replanted in a bed of damp soil. I peel mine and eat them cut in chunks dipped into curry-flavored mayonaise. (Raw cauliflower and Asian turnips are also good munching that way.)

Pests and diseases. Cabbage root maggots are a big problem with in-the-ground brassica crops in Cascadia and the United Kingdom. However, the rutabaga's thick skin seems to protect it; the skin gets scarred up, but the insides remain edible. The edible part of kohlrabi, forming as it does several inches above the soil line, is not invaded by maggots, and their presence in the roots does not do major damage during cool weather. Turnips and radishes, however, are nearly impossible in Cascadia unless you use fine-mesh row covers. I do not know enough to comment about problems in other climates. Where I live now there are no root maggot difficulties of any kind. Hallelujah!

Saving seed. All these bee-pollinated outcrossing crops suit home seed-saving because they are small plants; the gardener can maintain a large enough population to prevent inbreeding depression of vigor. All are members of the *Brassica* family (fortunately, they are different species that don't cross), and their seed is handled as described for kale. All require the closest selection for perfect shape or else your line will rapidly degenerate.

Salad radishes, if you recall, are annuals. Grow a large bed in spring and, out of 100 or so plants, select the most perfect 25 or 30 as you're harvesting for the table. Pinch off their larger leaves and replant them one per square foot (30 by 30 centimeters). Soon they'll be growing again and putting up seedstalks.

Winter radishes are intrinsically more vigorous. I have found that 10 or 15 plants is plenty for maintaining the gene pool's vigor. They must overwinter first. If cellared, replant in spring, giving each at least two square feet (1,850 square centimeters); four square feet (3,700 square centimeters) is better.

If you grow *turnips* primarily as a spring crop, let the last bolters (with good shape and other desirable qualities) of the spring sowing go on to make seed. If your interest in turnips is as an autumn/winter crop, grow seed from overwintered roots.

I have been saving seed of a local variety of *rutabaga* for four generations now, working from only a dozen plants in each generation, and have noticed no deterioration. Give each turnip, rutabaga, or kohlrabi at least four square feet (3,700 square centimeters) to make seed in.

Kohlrabi seed is produced like that of the other root crops. The vegetable is strongly biennial, and perfect specimens should be transplanted into a seed-making area after overwintering. If they overwintered in the garden, snip off at least half their leaf area (take the largest leaves) in early spring and set them in the earth right up to their crowns.

Cool-season greens: Spinach and mustard

When grown for autumn or winter harvest, consider these vegetables medium-demand crops. For spring harvest they are high-demand. Why the difference? Because they bolt in spring, so the plants must be pushed to harvestable size before going to seed — and that means the highest possible fertility, especially nitrogen. However, when sown for autumn or winter use, the plant will overwinter before bolting (or possibly freeze out), so there is no need to make the maximum growth at the fastest possible speed.

Growing details, spring. Both species will germinate in cool soil. In fact, where winter is severe, some gardeners sow spinach seed thickly in autumn with the reasonable hope that some of it will sprout shortly after the snow melts off. There can be many mysterious seedling disappearances in spring; sowing thickly and then thinning progressively insures against failure. Use strong compost or, better, COF made with some tankage or even one part highly potent bloodmeal instead of one part seedmeal. Work it into the bed before sowing, then side-dress closer to the rows of seedlings as soon as they start forming their first true leaf. Where I live, if the weather cooperates (meaning if the sun shines most days), the plants will get large enough before starting to bolt to make it seem worthwhile growing them.

Growing details, summer. Almost all varieties go to seed shortly after the seedlings begin growing. However, there are a few spinach varieties, bred for reluctance to bolt, that can be grown in summer. The seed catalog will proudly mention this trait. There is also one Asian mustard that will not bolt in summer. It is mild flavored, with broad off-white stalks, not as fancy as some of the tsi-tsoi types, but still pretty good. Johnny's Selected Seeds has a hybrid pac-choi for summer; Territorial still offers the older OP variety I have known and grown since the early 1980s.

Growing details, autumn/winter. Start mustard or spinach after the heat of summer has ended or a month after the solstice in short-season climates. They grow quickly; thin progressively.

Harvest. Except when thinning, do not cut whole plants unless they're starting to put up seedstalks. It's better to cut or break off large outer leaves. Do not take so many that the plant is crippled — one leaf per week from each plant in the row is about right. In spring, at the first signs of bolting, harvest the patch and resow the area to another crop. In autumn, if growing season remains but production nearly ceases, feed them.

Saving seed. Mustard is a bee-pollinated brassica. To maintain your variety's vigor, you must insure its gene pool contains at least a dozen plants; 50 would be better. Spinach is wind-pollinated and also has plants with gender: male, female, and hermaphrodite. The males only make pollen. Because roughly half of the seedmaking bed will be females, you need a starting population of at least 50 plants to maintain a spinach variety.

Close selection to a perfect type is not important in either species, however, any tendency to make seed ahead of the others is a highly undesirable trait. Early bolters should be destroyed before they can release any pollen. Spinach seed is generally made only from spring-sown plants, although it sometimes overwinters in mild-winter climates when not too much rain falls. It is not usually cold that kills it, but humidity. Mustard seed will mature from either spring-sown or autumn-sown plants (if they survive winter).

One advantage of growing your own seed is that it costs you nothing. In that case, consider using spinach as a spring-sown green manure. Early in spring, broadcast seed on roughly hoed beds, then cover it with broadcast manure or compost or shallowly hoe in the seed. The species makes a dense, tender root system that decomposes rapidly and leaves the earth in fine, friable condition. As soon as some of the plants start putting up seedstalks, chop the whole bed in with an ordinary sharp garden hoe. Wait a week for it to decompose; then, with a bit of raking, you'll have made an excellent seedbed. Mustards can be used as autumn green-manure crops where winter is severe. Start them at least 40 days before killing frosts.

Scallions (spring onions) and topsetting and potato onions

Straight-shanked onions are easier to grow than bulbing sorts. There is no need to push them into rapid growth and no need to sow them early in spring because scallions continue growing steadily until they either freeze out in winter or make seed the following spring.

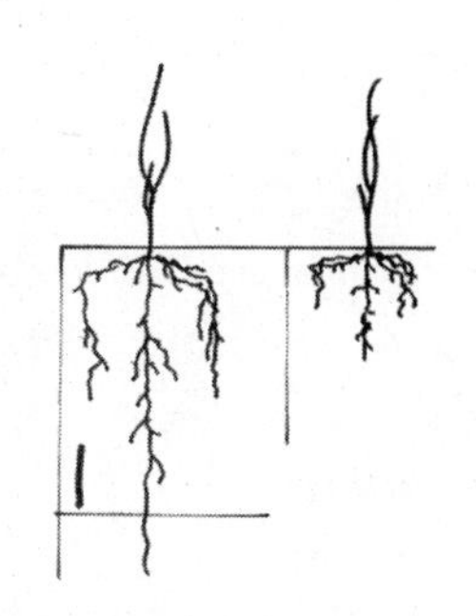

Figure 9.16: *Two onion seedlings: the one on the left grows in open, loose earth; the one on the right grows in heavy, compact ground. Both are drawn to the same scale. (Figure 8 in RDVC)*

Growing details. All onions prefer light, open soils, like loams or sands, and won't grow fast in heavy ground. If you garden on clay, prepare their beds with large quantities of compost or well-rotted manure in an effort to lighten it up. Digging a two-inch-thick (five-centimeter) layer into a clayey onion bed wouldn't be excessive. Scallions will eventually grow quite large if given only moderate fertility, but they can't grow at all well if they can't make roots.

I prefer to grow my scallions rather crowded and then harvest them by thinning. When the soil is naturally loose, individual plants can be gently tugged out of the row. That way, a single patch satisfies the kitchen's needs for the year. If you're gardening where summers get really hot and humid, it probably would be best to sow scallions after the main heat of summer has passed. Consider them a winter vegetable.

Pests and diseases. The minuscule sap-sucking thrip troubles commercial growers because it reduces the ultimate size of the bulb; rarely do home gardeners realize that there may be the odd one in their onions.

Onions of all sorts get fungal/mold diseases, especially in humid weather; these the gardener does notice. Molds and mildews can be serious trouble. Prevention is the best cure. First, give your plants more space so they get better air circulation. Judging only by their root systems, scallions could be grown on 12-inch (30-centimeter) between-row spacing, but I recommend 18 inches (45 centimeters) because it allows more air to flow around the plants. To enhance air circulation, thin early, thoroughly, and progressively. Small scallions can be finely minced as though they were chives. Second, after harvest, clean up all onion trash and compost it; this action provides fewer overwintering havens for disease organisms.

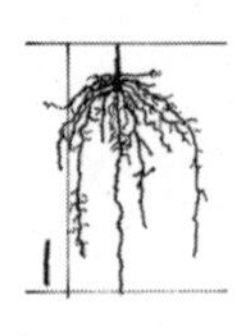

Figure 9.17: *A two-month-old onion seedling in good soil. Notice the minimal root development. (Figure 7 in RDVC)*

Onion molds can invade the vascular system. Once they are inside the plant, only systemic fungicides of the sorts commercial growers use will save the crop. It is possible that frequent spraying (every few days) with mild external fungicides like lime sulphur or compost tea will prevent infection. I have found that leeks and garlic have far higher resistance to these diseases then onions do.

Varieties. Most seed companies these days sell only *Allium fistulosum* varieties, also known as Welsh onions. However, Stokes still offers my favorite, the Lisbon onion, which is *Allium cepa* and will cross with other bulbing onions, which are also all *A. cepa.* Why do I prefer Lisbon? Because, although smaller and more trouble to prepare for the table, it is mild, tender, and sweet, while Welsh onions tend to be a bit hotter and tougher. Lisbons, however, are not very mold resistant.

Topsetting onions are a less-refined and usually easier to grow medium-demand crop. They are a non-bulbing perennial, much like a giant chive. But instead of forming a seedmaking flower, they form little bulblets on top. The weight of maturing bulblets makes the stalk fall over. Then the bulblets self-sow a foot (30 centimeters) or so away from the plant and start another plant. That's why this sort is sometimes called "walking onions." Harvest the stalks one by one as needed before the bulblets form. Unmanaged topsetting onions that walk too far can become a pest. To grow them, plant a few bulblets begged from a neighbor, or a bit of a root crown.

Potato onions are actually a type of shallot with medium-demand fertility requirements. They have two uses: the bulbs can be used as rather pungent cooking onions that store well; the tops taste good, make an acceptable substitute for scallions, and tend to be far more mold resistant than the usual onion variety. To grow them, get one bulb (or a few) and plant it shallowly in spring (or in autumn in climates with mild winters). Snip the odd stalk as needed for scallions. Otherwise, treat it as though you were growing garlic. This is an excellent alternative for someone having difficulty growing the usual bulbing onions.

Saving seed. Saving seed from scallions entirely suits the gardener because 25 or more plants don't use much space. If a bed of scallions will overwinter where you garden, it'll go to seed in spring. Allow the seedballs to mature on the plants and harvest them before the small black seeds start falling out of their capsules. Watch each plant; when you see a few mature seeds showing in one, tug the stalk out of the ground, place it on a tarp in the shade under cover, and allow it to continue ripening and drying. When the stalk has become brown and crisp, the seed should be totally done. Then clip off the seedballs and put them in a large shallow bowl; bring the bowl into the house and put it in a warm place until the seedheads themselves are crisp enough that when you rub them between your hands, the seeds will pop out of their capsules. After rubbing, winnow in a light breeze. You may find this surprising, but your own seed, allowed to fully ripen this slow and patient way, will usually be superior to anything bought commercially. Alliums are bee-pollinated and need about half a mile (800 meters) of isolation from other onion flowers in the same species (as mentioned earlier, there are two onion species).

Garlic

I produce about 50 heads of garlic each year that hang in braids decorating our kitchen, and another 30 to give away as gifts.

Growing details. Garlic overwinters as a small plant. In spring it rapidly grows leaves, the leaves make surplus food that is stored in belowground bulbs, and then the plant goes dormant. In every variety I've ever grown successfully, bulbing happens while days are long and still lengthening. It is probably different with varieties suited to tropical latitudes.

To plant garlic, first break the heads up into individual cloves. One clove becomes one plant that forms one head containing many cloves. If you use only the larger cloves for seed, you'll harvest larger heads. Plant cloves root-side down, about two inches (five centimeters) deep. If the plants bulb in compacted soil, the heads will be small, so if you have heavy soil, lighten it up as though you were growing scallions (or any other allium) by blending in a goodly amount of compost or well-aged manure before planting.

Most varieties are harvested before the summer solstice; a few very late ones are dug shortly after that. The size of the head is completely determined by the size the top has achieved before lengthening days trigger bulbing. If you want big heads, you have to push the tops for size during the months this can be done — give them fertile soil.

Garlic may be planted early in spring, but the heads will form somewhat later and will be considerably smaller. Even where the soil freezes, garlic should be planted to overwinter unless that proves impossible where you live. David Ronniger, from snowy northern Idaho, plants his garlic late in September. He says that if the little plants are given some mulch before the snow falls, they survive winter and start growing again first thing in spring, with a big head start on anything sown in the spring. I sow mine (in a maritime climate) about a month after the equinox.

Garlic produced in a climate that allows the tops to grow vigorously over the entire cool season (in California, for example) gives the highest yield. If that climate features a predictably dry spring, there is no risk of rots occurring in the ground, and the crop will be of the highest quality as well as the highest yield. That is why almost all North American commercial garlic is grown around Gilroy, California.

As soon as growth resumes in the spring (relatives of the alliums such as daffodils and crocus will come up at this time), I suggest side-dressing the garlic patch with high-nitrogen fertilizer, COF, or chicken manure compost. Keep it well weeded; the shallow roots are not able to compete. Your entire intention should be to grow the largest possible top before bulbing begins.

Exactly when bulbing starts depends on the variety. A month before bulbing starts, there is no longer any point doing more fertilizing. From the time bulbing first begins, dig a plant once a week to learn by observation how your variety matures. If you are routinely irrigating, keep in mind you'll harvest higher-quality garlic if it is dug from soil that is just barely damp.

Pests and diseases. Insects and disease are usually not a problem above ground, but rots will happen if the soil is too moist when bulbing is close to finishing. There are virus diseases that infect garlic and carry over from year to year in the cloves. Most of these are called "yellows" because they cause premature yellowing of the leaves and considerably reduce the yield. It's to your advantage to start out with disease-free seed.

Varieties. There are many varieties of garlic. Some only suit tropical daylength patterns; most varieties available in seed catalogs are bred for temperate latitudes. In Australia, much garlic for eating is imported from tropical China; these varieties don't perform well farther from the equator. I suggest visiting ethnic specialty markets (Greeks and Italians use a lot of garlic in their cooking) and see if you can find some locally grown garlic for sale; use that for planting stock. This will save you quite a bit of money compared to buying "seed" garlic.

There are two basic types of garlic: hardnecks and softnecks. When hardnecks start bulbing, they also put up a central stalk crowned with tiny bulblets. Mature bulblets may be planted; each will make a small garlic plant that makes a small head the first year. Often this first-year head contains only one large round clove. Unless your intention is to rapidly increase your planting stock for the purpose of selling garlic, it is best to clip off that seedstalk as soon as it appears. This allows the plant to put its entire resources into forming cloves below ground. Hardnecks are difficult to plait because the stalk remains.

Softnecks do not make this secondary set of bulblets. The entire top shrivels as onion leaves do, so they are easier to plait. Softnecks also tend to store longer and have slightly milder flavors (but not always). However, their cloves tend to be a bit smaller. I grow one hardneck and two softneck varieties.

Finally, there is a non-garlic garlic, called Elephant garlic. It actually is a giant leek with a garlicky flavor. It also puts up a seedstalk, but one that makes seeds. Gourmets disdain Elephant garlic, although in terms of bulk yield it may be the highest-producing form. There is a serious downside to Elephant garlic; as many varieties of leeks do when making seed, it forms little corms at the base of the bulbs (gladiolus bulbs also do this). Inevitably, some corms are overlooked at harvest and resprout. It can be nearly impossible to eliminate this crop from a bed it grew in without more than a year of constant hoeing.

At first look, growing garlic seems an attractive and highly profitable homestead enterprise, needing little land to make a big income. But there is an obstacle — the labor crunch that comes at the time the heads must be peeled and dried. There are never enough hands in the family to process nearly as much garlic as there is labor in the family to grow. There is also a lot of risk should the weather go against you and turn rainy just before harvest. So before you invest huge amounts of money (for seed), time, and energy into this as a business venture, I suggest you first figure out how to grow three or four different varieties whose maturity is spread out over enough weeks that you can process the harvest. Then increase your own planting stock for a few years while you work out the peeling and drying of large quantities.

Harvest. Dig the plants when the cloves have fully segmented but before the heads begin to burst open or otherwise deteriorate (dig one every week and learn your variety's behavior). Gently shake off the soil. Bursting heads permit soil to enter; these heads won't ever store well, and they don't look attractive, either. Garlic dug before maturity makes excellent cooking stock; because its flavor is milder, it is especially good in salads. Garlic heads dug before segmentation has started are like small garlic-flavored onions, excellent in stir-fries. When it's time for the main harvest, dig all remaining plants and carefully spread them out under cover in the shade with good air circulation. Dry the plants for two or three days and then strip off two or three layers of leaves. Peeling removes all clinging soil and exposes the beauty of the bulbs. Then allow drying to continue (it's best to dry them slowly) until the leaves are dry enough to plait, or else let them completely shrivel and clip them off an inch (2.5 centimeters) or so above the bulb. Some growers who braid garlic allow the leaves to shrivel almost completely and then remoisten them, softening them up enough to braid. If you don't hang braids in the kitchen, put the heads in an onion sack and hang that; lots of air circulation is crucial to long storage.

Saving seed. A portion of your harvest is your seed stock. I always reserve the appropriate number of my most perfect heads for planting stock.

Difficult vegetables

The vegetables in this group require abundant soil moisture throughout their growth and/or require extremely fertile soil and/or have other tricky aspects. Asparagus is here because starting it is an investment taking several years to pay off, and it also demands good management if the bed is not to be ruined.

Onions, bulbing

Let me warn you in advance: This section is complex because I am not a Pollyanna garden writer. I'm an ex-seedsman who wants his friends to grasp the real story.

Growing details. If this weakly rooted crop is to thrive, not just survive, it needs loose, fertile, constantly moist soil; minimal competition from other onions nearby; and no weeds to contend with. The species is disastrously prone to mildew-like diseases, which is another reason not to crowd the onion bed. These diseases rarely cause trouble when the crop grows in dry air with enough elbow room to allow good air circulation. Humidity is the prime reason bulbing onions are not grown during the summer in the southern United States. Mildew is the reason that onion growers on Tasmania use fields exposed to almost-constant sea breezes.

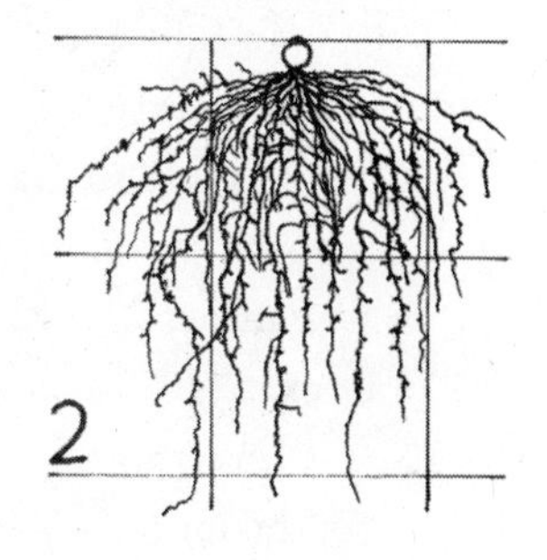

Figure 9.18: *A long-day storage onion plant 105 days after direct seeding. Bulbing is half completed. Little additional root development will happen from this point. (Figure 9 in RDVC)*

The onion grows vegetatively until a certain date (based on daylength at the latitude it is being grown at); then it forms a bulb whose size depends almost entirely on the size the tops achieved before the start of bulbing. The changeover to bulbing has little to do with how long the plant has grown. Usually if you have achieved tall husky tops before bulbing starts, you'll get big bulbs. Getting a big top is especially important when growing sweet intermediate-day onions because if these types don't get big, they won't be sweet.

There are three basic sorts of onions used in temperate climates:

- *Short-day varieties* are bred to be grown at latitudes below 35°, where winter is mild enough that they don't freeze out. Short-day varieties are sown during the cool season (autumn/winter), and overwinter in the field as small plants. They finish bulbing before the summer solstice and are usually mild, tender, and sweet. Rarely are they long-keeping onions. A few varieties are hardy enough to overwinter in a maritime climate.

- *Intermediate-day varieties* are intended for growing between 32° and 38° latitude. They are sown in the spring and tend to be on the sweet side.
- *Long-day varieties* are intended for growing from 38° latitude on, as far toward the pole as agriculture can be done. They are sown in the spring and tend to be hard, pungent, and long-keeping, exactly what is needed in a climate with a severe winter.

To fully comprehend using the three types of onions, you need to understand how daylengths work at different latitudes. (The full explanation of this belongs in a geography text, not a garden book.) To tickle your memory, let me remind you that on the two equinoxes, which occur around March 21 and September 21, every place on earth has a day of exactly 12 hours and a night of exactly 12 hours duration. The longest and shortest days occur on June 21 and December 21. One more crucial thing: the farther from the equator you go, the longer the longest day is; the shorter, the shortest day is. For example, at 45° latitude the longest day is exactly 17 hours from first light to last light, while at 35° latitude the longest day will be only 15 hours, 35 minutes.

Suppose we breed an onion variety that will start bulbing when the daylength is both decreasing and has reached 15 hours and 34 minutes. If grown at 35° latitude, that onion will begin bulbing one day after the solstice. If grown at 45° latitude, that same onion will begin bulbing either February 5 or August 5, depending if you're in the northern or southern hemisphere. August 5 might be a perfectly appropriate date for the start of bulbing of a storage variety grown at 45°N latitude, leaving the bulb a month or more of warm weather to finish sizing up and be cured, but that same variety, grown at 35°N, hardly has a chance to take advantage of the summer to grow a big top before bulbing starts, resulting in a small onion that matured far too early to keep through the winter. A spring-planted onion bred for the mid-latitudes (i.e., bred to start bulbing at about 14 hours of daylength) would grow vegetatively until mid-August. But if an onion that started bulbing at 14 hours of daylength were grown at 45°N, it wouldn't start bulbing until August 28 (or February 28 in the southern hemisphere). This would result in maturity being so late that autumn would have arrived before the onion had finished growing — and you can't cure an onion in cool, humid conditions.

Please thoroughly consider the preceding paragraph; read it over a few times until you fully grasp the concept. Then it should be clear to you why I say that the main tricks to growing large-sized intermediate-day and long-day onions are: (1) to use a variety that matches your latitude, and (2) to direct-seed them as early as possible in spring (or even late in winter in a greenhouse or heated frame) and then supply conditions that will help them grow as big as possible before they must bulb. Fortunately, onions transplant easily, especially in cool weather. Gardeners seeking to produce lunkers often grow their seedlings in flats or trays, putting them out in the garden as soon as spring weather settles down a bit. When directly seeded as early as possible, bulbing varieties won't achieve "lunker" class, but still make a respectable harvest if they otherwise get what they need in terms of fertility, soil conditions, and moisture.

So what about short-day varieties? These grow vegetatively during that part of the year when the days are shorter, and bulb when daylength increases. They are the sweetest, mildest sort and are my favorite type for eating raw with bits of aged cheese and a few olives, or sliced into salads.

The onion is frost-hardy, but not infinitely so; tiny seedlings are not as hardy as plants with three or four leaves. So short-day onions are usually grown where winters are only a bit frosty at worst. Of all the sorts of bulbing onions, these are the trickiest because if the plant should happen to grow too well during an unusually mild winter, if it should get too big too soon, the plants will fail to make bulbs and will, instead, produce seedstalks (called "bolters"; bolters are useless for the table). The key to success with them is to divine the correct sowing date. The formation of seedheads (instead of bulbs) is triggered when their growth stops and resumes several times over winter after the plant reaches a minimum number of leaves. In simple terms, if the plant has not gotten larger than a lead pencil when its growth is checked in winter, it will still make bulbs. If its stalk is thicker than a lead pencil and harsh winter weather checks its growth, it will almost certainly make seed and not bulbs. This means you should sow onions later in the autumn rather than earlier: early enough that the soil is still so warm that the seeds will sprout and the plant will grow a few leaves, will become big enough to withstand a bit of frost, but not so early that it will grow big before spring and become a seed crop. It also means it is wise not to make the soil super-fertile when sowing short-day onions. Instead, you should side-dress the crop heavily as soon as it has securely resumed growth in spring — i.e., when the ornamental bulbs like crocus or daffodil, onion's relatives, have begun coming up.

One special case with short-day varieties: Some have been bred to be extraordinarily cold-hardy. These can be grown over winter in maritime climates.

Gardeners living on the boundary between short-day and medium-day territory who have a greenhouse and the desire to grow prizewinning lunkers may start both short-day and medium-day varieties in flats about the time of the shortest day of the year, grow them as fast as possible under protection, and then transplant them out to their beds as soon as conditions settle. Gardeners who live where winter is severe and who have a heated greenhouse or hotbed may start medium- or long-day varieties during winter between the solstice and the equinox, grow them as fast as possible in a greenhouse or frame, and transplant them out as early as possible. However, growing lunkers is not necessary. With far less trouble you'll harvest nearly as many pounds per square foot growing medium-sized bulbs on somewhat denser spacing.

Remember that bulbing onions, like almost all other alliums, require extremely fertile, airy, always moist soil. The usual procedure for all types is to sow seeds thinly in furrows and thin progressively so that by the time the seedlings have achieved the diameter of lead pencils, they stand as far apart in the furrow as you hope they'll be at maturity — usually three to four inches (seven to ten centimeters) apart unless you're growing a small-bulb variety. Keep the crop well-irrigated until bulbing is about half completed and then hope the soil will be rather dry at the time the shriveling tops fall over. There is no point in side-dressing or foliar feeding after (or even a few weeks before) bulbing begins. Otherwise, grow them like scallions.

Pests and diseases. See the information for scallions.

Varieties. As hybrids have taken over all commercial applications, the classic OP sorts, lacking the loving attention of plant breeders, have degenerated into something nonproductive. (Translation of my polite phrases into plain English: Do not buy cheap seed.) One exception to this concerns only a half million Tasmanians, whose island has an industry growing long-day onions of an OP variety called Creamgold; seed stocks for Creamgold are still excellent. Make sure you know if you're dealing with short-, medium-, or long-day varieties and grow them accordingly.

Harvest and storage. All sorts may be eaten as scallions (spring onions) before bulbing begins. When about half the tops have fallen over, push down the rest. Then wait a few days and pull them from the earth. It might help to loosen the soil a bit with a shovel first. Ideally, let them lie on dry soil in the sun to cure for a few days. If the soil isn't dry or the weather isn't settled, bring them indoors, spread them on a tarp under cover, and allow them to dry until the tops have completely shriveled. Onions in storage need lots air circulation and cool, dry conditions. I suggest begging some onion sacks from the local supermarket and storing your harvest in these bags, hanging them from strong hooks.

Saving seed. In maritime and other mild-winter climates, please do not produce onion seed by sowing seed late in summer. This shortcut does not permit you to select perfect bulbs and reject off-types; in the seed trade, this production method is called "seed-to-seed" and is only done by primary growers of the cheapest home-garden seed, unethical businesspeople who do not care if their product actually makes many bulbs. Always grow onion seed by planting mature, hand-selected bulbs. Choose well-wrapped, middle-sized bulbs with the correct shape and narrow necks (which usually result in longer storage). You should include at least 50 bulbs in the gene pool. Where winter is severe, plant the bulbs out in spring after wintering them over; set them about one foot (30 centimeters) apart in rows two feet (60 centimeters) apart. Plan on staking up or otherwise propping up the seedstalks. Where winter is mild, bulbs may be set out in autumn (or late in summer if they're sprouting). Otherwise, grow the seed as though for scallions.

Leeks

Depending on your climate, raising this slow-growing vegetable can be easy or daunting. In a maritime climate, leeks effortlessly withstand winter, sizing up during mild periods and holding fast when the weather turns really chilly. In this circumstance, leeks are not difficult because you have lots of time to wait for them to get big. Where autumn rapidly turns freezing, leeks must be pushed to make the most rapid possible growth and then harvested before it gets too cold. This means you must use techniques that allow for the earliest possible sowing and provide ideal soil conditions throughout their seemingly endless growing period. Where summers get too hot for them (and for most other members of the onion family), leeks must be started after the worst heat of summer breaks. This gives them little time to size up because in all three situations the leek, having overwintered, will go to seed shortly after the spring equinox. So in the American south or in the warmer parts Down Under, growing leeks also means sowing after the end of hot weather and then pushing them hard with maximum fertility and soil moisture.

Growing details. Leeks take 150 to 180 days to achieve admirable size. I wouldn't intentionally grow leeks to eat when they are small; in that case it would be easier to grow scallions (spring onions). In all climates, start them as soon as possible in a carefully prepared nursery bed where the seedlings can be given ideal conditions. I find that a single four-foot-long (120-centimeter) row of seedlings produces enough to transplant for my garden and the gardens of two friends. Take a small bit of bed or row and amend it with at least two inches (five centimeters) of your best compost or well-aged manure, well mixed in. Also give it COF or chicken manure compost or some other rich source of complete plant nutrients that becomes rapidly available to the plants. Then sow the seeds rather densely, say, 12 seeds to the inch (2.5 centimeters). In a maritime climate, do this when convenient in spring after the weather has settled; about when the apple trees are blooming.

In mild-winter/hot-summer climates, start leeks on the earliest possible date after the main heat of summer that you can be sure to get germination. It would be wise to presprout the seeds. Consider putting the nursery row under a structure that provides about 40 percent shade for the first month.

Now, GROW those seedlings. If you get a good germination, thin them to about six to the inch (2.5 centimeters) after a month or so. Side-dress the seedlings with a bit of extra fertilizer if they aren't growing really fast (for leeks). If you can manage it, don't ever let this little bit of row get dry. In my maritime climate, depending on variety, it takes leeks about 90 days to get to the size of lead pencils. That is transplanting size and is also about when they are getting far too crowded on a spacing of six to the inch. Leeks are a special case in terms of competition; it is desirable for them to compete in moderation because struggling against each other makes them "leggy," creating a larger gap between the roots and the place on the stem where the first leaf emerges. This elongation is desirable, as you'll soon read.

Immediately before transplanting, dig up the entire seedling row, gently shaking the earth from their roots while doing as little damage to the roots as possible. Now separate them by size (large, medium, and small) and closely inspect them. If you're planning on growing your own leek seed, sort them again within the medium-sized category and remove any that show a swelling at the bottom. Put those without a perfectly straight shank back with the large ones. These big ones and nonperfect medium-sized ones are the seedlings you'll replant to grow for the table. The medium-sized ones with straight shanks and the longest space between root and first leaf are the ones to grow on for a seed crop. (If you're growing seed, make sure there are at least 25 in this lot.) If you have neighbors to whom you can give leek seedlings, take the small ones, replant them shallowly in the leek nursery, spaced at about half an inch (1.25 centimeters) apart, and grow them on like scallions for another month or so before digging them again and giving them away.

Now, grasp the big and medium-sized seedlings in large handfuls, held so that their roots are all at the same position, and with a sharp knife cut off approximately half the leafy part (not including the length of their stem); with less leaf they won't have to draw so much water when they are put back into the earth. This insures 100 percent success with transplanting. (Do this with the rest of your seedlings, too — the ones for gifts and those for making seed.) Keep bare-root seedlings out of the sun. I put them in a bucket, roots down, with about a quarter inch (six millimeters) of water on the bottom of the bucket. This will hold them while you prepare a place to transplant them.

When preparing soil for the "for eating" transplants, dig it deep. Make their bed as loose and as fertile as you possibly can. With a pointed stick about 1½ inches (four centimeters) in diameter, poke a hole in the soil about eight inches (20 centimeters) deep and large enough around that the seedling (and its roots) may be dropped in. Bottom out the seedling and then draw it back up so that the roots are stretched out, pointing downward. When the first leaf that emerges from the stalk is at the same level as the soil's surface, push in earth to fill the hole and hold that seedling in place. The part that is underground will make a white shank, the most desirable part to eat. (This is why you grew the seedlings to make them as leggy as possible.) Side-dress a bit of fertilizer along those rows in another month or so, and if you can, pull up a bit of earth against the stalks with your hoe as they grow. Only do this when you can avoid getting soil into the lower leaf notches; soil in the leaf notches will be taken into the kitchen at harvest. Hilling up more earth increases the length of the white shank.

Pests and diseases. Pests and diseases should not be a problem if you do not grow leeks (or other alliums) in the same soil more than once every three or four years.

Varieties. There are three kinds: autumn, winter, and spring. Autumn leeks grow faster and are less cold-hardy. They are intended for commercial production in Europe, to supply the market before the weather turns too cold. Their flavor is usually milder; their texture more tender. Use these as a cool-season crop in climates where summers are too hot for leeks.

Winter leeks grow a bit more slowly and are tougher, more fibrous, more pungent, and much hardier. Generally this is the sort offered in home-garden seed catalogs.

Spring leeks are bred to bolt as late as possible; they are intended to supply the European spring market after overwintering in the field. Since I garden in a maritime climate similar to Holland, Belgium, or the north of France I prefer "spring" varieties, but you may not be able to obtain these outside of European seed catalogs.

In every type of leek, good breeding shows as a non-bulbous stalk that is long to the first leaf notch.

Harvest. Leeks must be dug. They are remarkably cold-hardy and even in places as cold as Pennsylvania (at low elevations), the hardiest varieties may survive winter in the earth if some soil is pulled up against them. But they are not infinitely able to withstand long freezing. Leeks may also be held over winter in the root cellar.

Making seed. Make seed as you would do with other alliums. In my mild-winter climate, I transplant carefully chosen perfect specimens in a separate area, and don't bother to bury them deeply nor to make their soil super-fertile. Where the soil freezes in winter, replant them about one foot (30 centimeters) apart in rows about two feet (60 centimeters) apart when they come out of the cellar after overwintering; there's no need to replant these deeply. Otherwise treat them as though you were growing seed for giant scallions.

Leeks won't cross with ordinary onions. The seedheads sit atop enormously tall stalks and usually have to be supported. Seed maturation seems to take endless time. Eventually, even though the small black seeds are not yet showing inside their capsules, you'll find that you can pull the stalks out of the earth with the gentlest of tugs. Each seedstalk loses its roots at a different point. Test each one about once a week. When a stalk lifts from the bed easily, place it on (or hang it upside down over) a big tarp to catch some of the seeds. Do this under cover with good ventilation and let it continue drying down. At this stage the seed is still forming, being fed by the nutrition and moisture in the stalk. Eventually the stem will dry out below the seedball. At this point I cut the seedballs free from their stalks and bring them inside, where it is warm, to finish drying to a crisp before I rub them between my palms, releasing the seeds. In my climate it takes until mid-autumn before the seedballs are crisply dry. Where summer is hotter, I expect it would go faster.

Chinese cabbage

If you want a large, tight, succulent head to form before this vegetable puts up a seedstalk, then you need (1) top-quality seed, (2) highly fertile soil that is always moist, and (3) knowledge about when to plant.

Growing details. Chinese cabbage is a close relative of mustard, and like mustard it has an almost irresistible tendency to go to seed when days are long. Breeders have not yet fully overcome that trait. Sowing early in spring for harvest in early summer is extremely dubious, even if you have purchased a variety that pretends to this ability. Sowing in late spring for harvest midsummer is nearly as dubious. My advice: Do not start this crop until after the summer solstice. It's best to wait at least one month after that event. In mild-winter climates wait two months. Timing is everything. In warm climates where the crop may be grown to mature in the cool months, it must still be sown in time to head up before spring days get too long. You should certainly sow it in time to be harvested around the equinox. In colder places, if you start it too late, growth will slow with the onset of chilly conditions, and the plant may put up a seedstalk before it heads, or the head may be tiny and of poor quality, or it may freeze out before doing much of anything. You can only determine the right sowing date by trial and error. If you're eager to attempt this crop, I suggest that the first time you do so, you sow two heads every two weeks over a few months. See what happens.

Speed of growth is everything with Chinese cabbage! Make the soil as rich as you know how. Keep it moist. Don't crowd the crop or competition will slow growth. Each plant needs to control about four square feet (3,500 square centimeters). Put four seeds in the bottom of a thumbprint atop an especially well-fertilized spot. Thin to a single plant as they begin to compete. Certainly within one month of emergence there should be a single one. Do not try to transplant Chinese cabbage; it forms a taproot and never grows as well without one. Keep that in mind if considering purchase of garden center seedlings.

Pests and diseases. All the usual brassica pests may pester it. All the usual pest remedies apply. In Cascadia, the cabbage maggot has an especially nasty way of feeding on Chinese cabbage: the maggots tunnel across the bases of the leaves, collapsing them and rotting the plant. When I lived in Cascadia, I grew Chinese cabbage under a spun fabric tent to exclude the fly.

Varieties. I know of no OP varieties that retain much uniformity; if you grow an OP, be prepared in advance to have over half your plants fail to head before bolting. If you use a quality hybrid, be prepared to have nearly every plant make a perfect head. There is a non-heading variety called Santoh, which compares to Chinese cabbage as collards do to European cabbage. Santoh is much easier to grow.

Harvest. If, when harvesting, you do not cut off the entire base, but instead allow a few of the larger outer leaves to remain attached to one side of the stem, the plant will continue putting up small leaves and then seedstalks; both are good in salads or stir-fries. Thus a non-heading plant or early bolting is not a complete loss if you remember that the upper portion of seedstalks are good in stir-fries if cut before the flowers open.

Asparagus

Asparagus is a perennial whose establishment most people believe involves a great deal of effort and some considerable expense. They also believe they should not harvest it for the first two years after establishing a bed. They learned this from Everybody Else.

Almost all people defer to the viewpoint of Everybody Else before deciding what to do about something. About growing asparagus, Everybody Else states confidently that two-year-old crowns should be transplanted into ditches that have been deeply dug and made super-fertile. Naturally, if Everybody Else is in favor of it, I will come to disagree. It always seems to work that way in my life.

My method will give you a much more productive result at a far lower cost and in far less time. If you find the difference between my viewpoint and that of Everybody Else distressing, keep in mind that wild asparagus doesn't transplant itself. Its seeds merely fall to earth, sprout, take hold, and survive freezing winters. Also remember that the reason asparagus beds peter out is because overcompetition from seedling plants started from seed made by the bed itself reduces the size of the shoots.

Growing details. You can grow asparagus if you live where there is sufficient winter chilling for it. That doesn't mean freezing soil. Just a few cool months will do. In North America that generally means north of southern Georgia, and certainly anywhere south of Sydney in eastern Australia will be okay.

Asparagus does not do well in clay. Period. In times of heavy rain it requires well-drained soil or its roots become diseased. If clay describes your situation and having asparagus is your passion, consider making a special bed of sandy soil that is raised up at least one foot (30 centimeters) above everything else for planting it. I knew a gardener who grew excellent asparagus on a clayey site that was occasionally underwater during winter rains. He built a marine plywood box that was four feet (120 centimeters) high, four feet wide, and eight feet (240 centimeters) long, filled it with sand dug from beside a nearby creek, and grew his asparagus in that.

In early spring prepare a four-foot-wide (120-centimeter) raised bed as though for growing a high-demand crop. Asparagus ferns can exceed five feet (150 centimeters) high and may shade neighboring beds, so consider that when choosing its position. After making the entire bed fertile and digging it well, make an extra fertile strip down the center of that bed. At minimum, spread a layer of compost or aged manure about an inch (2.5 centimeters) thick and a foot (30 centimeters) wide, running down the center the long way. I would also spread about three quarts (three liters) of COF for every 25 feet (8 meters) of row. Mix these amendments in thoroughly, digging as deeply as your shovel will go. Make a furrow about half an inch (1.25 centimeters) deep down the center of that extra-fertile band of soil and sow asparagus seeds two inches (five centimeters) apart. Do this about the time the apple trees are blooming, before it starts getting hot. Keep the seeds moist until they sprout.

A month after the seedlings come up, thin them to about six inches (15 centimeters) apart and side-dress them with some chicken manure compost or more COF. If you want to harvest a little asparagus the next year, *make them grow fast in their first season.* It is also a good idea to irrigate these seedlings every few weeks during their first year's growth. In subsequent years the bed should survive long periods without rain without problems.

Keep the bed thoroughly weeded. Always. In the beginning and in coming years, too. The need for a weed-free bed is the hardest part of attaining success with asparagus. If your bed grew well, the plants will make seed toward the end of summer. Some plants will prove to be males, which form little sacks that release pollen. The females make seedballs that turn red when ripe. Before the seedballs ripen, and before any seed drops on the bed, dig up and destroy the root crown of all the female plants. Make absolutely sure they are all dead. (When you dig these six-month-old female plants, you'll be amazed to discover that the roots you are excavating with such difficulty have already become huge, far larger than so-called two-year-old crowns for sale at your local nursery.)

Now you have an all-male row that, by the law of averages, will be spaced roughly a foot (30 centimeters) apart. If there should be any gaps that are more than 24 inches (60 centimeters) wide, sow a few more seeds in these spaces the next spring and again remove any that prove to be females. If the next year a few more plants turn out to be females, dig them up too and destroy them. (You may not have identified them because they did not make seed their first year.)

Why create an all-male bed? Because seed formation is a huge burden on the plant and reduces the size of its spears. Making pollen is not nearly so much effort, so males make more shoots and larger ones, which is what you want. This is why plant breeders have lately come out with all-male hybrid varieties. There's another, more important, reason for an all-male bed. Everybody Else confidently tells you that an asparagus bed lasts only 10 to 15 years. That is not necessarily true, as is the case with most of what Everybody Else tells people. Asparagus beds peter out mainly because of self-sown seeds.

These cause the density of the planting to increase, and this incredible over-crowding wrecks the bed. But if no seed drops on the bed, the bed continues producing large handsome spears — as long as it is kept fertile and weeded.

Ongoing maintenance. Each spring before the asparagus starts shooting, cover the bed with a half-inch-thick (1.25-centimeter) layer of finished compost or aged manure and a hefty dose of COF or other complete fertilizer. Asparagus doesn't absolutely have to grow in super-fertile soil, but it does if you want a big harvest. If you don't use COF, at least give it a dusting of lime (five pounds per 100 square feet of bed or 2.25 kilograms per ten square meters) once a year. And keep the weeds out. Thoroughly. The crowns will gradually spread out across the entire bed. One other thing you might do if you don't garden right next to the ocean: each spring before the crowns shoot, you could broadcast half a pound of rock salt per square yard (250 grams per square meter) of bed. All the 19th-century growers swore by salt on their asparagus beds. Those living by the ocean might mulch their beds with seaweed, which will naturally provide the salt.

Pests and diseases. The worst insect pest is the asparagus beetle. It, and its larvae, defoliate the ferns, reducing harvest in the coming year. If there are many, spray them. Pyrethrum/rotenone will serve if you use it every few days when the beetles are active. The most important preventative should be done during early autumn. Clip all ferns at ground level (in a manner that won't damage the crowns) and remove them from the bed. If you don't immediately compost the ferns in a heap hot enough to kill the beetles and their pupae, then burn the ferns.

Varieties. Gardeners using all-male hybrid varieties have reported disease problems; the hybrids that produce the usual 50/50 mix of sexes are fine. Unless your local agricultural experts have a strong recommendation, my suggestion is to start with UC 157, which is commonly available, is not expensive, and has proved adaptable to most places.

Harvest. It is possible that your asparagus plants will begin making shoots larger than three eighths of an inch (nine millimeters) in diameter as soon as their second year. Harvest all shoots this size or larger. The rule for harvesting is the same no matter how old the bed is: Any shoot that exceeds three eighths of an inch in diameter may be eaten; any shoot smaller than that is allowed to continue growing and to form a fern that recharges the root with food. The first flush of shoots in spring will usually be the largest ones. As you clip them, the stored food in the roots will gradually become depleted and the shoots will get smaller. As soon as there are no more shoots exceeding three eighths of an inch, it is time to stop harvesting. In the second year of a bed's life, the harvest may go on for only two weeks. In later years it might continue for six weeks or more. Once you've quit for the year, should any shoots come up in summer that are large enough to harvest, it is best not to cut them. Give the bed every possible chance to recharge its food storage for the next spring's production.

Saving seed. My advice: Don't allow seed to happen. To grow all the asparagus you'll ever want for as long as you garden in a place, you'll need to buy only one big packet of seed.

Celery and celeriac

Celery (and celeriac) require deep sandy loam soil to grow to perfection. Sand doesn't hold quite enough moisture and the result will show that; soils containing more than about 25 percent clay will not grow great celery. Clay soils (more than half clay) are out of the question. I know that nearly every clay-soil gardener is going to try growing celery anyway; you folks should know that there is an herbal form, far better adapted to poor situations. It is called "Chinese" celery or "cutting" celery and is a short-stalked type with a stronger flavor, used wherever regular celery might be included in a recipe.

Growing details. Celery and celeriac require maximally rich soil that unceasingly provides readily available moisture. The bed must also be worked deeply because celery makes a rather weak taproot that needs to extend downward over four feet (120 centimeters) if the plant is to grow properly. Wild celery grows in marshes; because it has little need to forage for moisture, its lateral root spread might not extend much more than a foot (30 centimeters) from its center. This is one crop for which you should double-dig. To prepare the celery row, simply dig a ditch one shovel's blade deep; temporarily place the soil beside the ditch. Put a two-inch-deep (five-centimeter) layer of well-aged manure or fine compost into the bottom of the ditch, as well as a full dose of COF (or a goodly dose of chicken manure compost and a light sprinkling of lime), and then dig that into the subsoil, going down another shovel's blade deep (and wide). Then spread a similar amount of manure, compost, and fertilizer over the soil removed from the trench so that when you refill the trench, the amendment will be mixed into the topsoil. Now the celery can grow in a zone of super-fertile loose soil about 12 inches (30 centimeters) wide and at least 20 inches (50 centimeters) deep. If, when going that second shovel's depth, you encounter clay and you hope for good celery, remove the clay and fill the ditch with topsoil. I have to do that. Blend the clay you removed into your next compost heap.

Celery grows extremely slowly. It can take ten weeks, easily ten, to reach the size that most people would consider a "transplant." So if you're growing a large celery patch and direct-seed it, you may be wasting a lot of valuable growing room. I suggest that you start a nursery bed instead. Make it as fertile as the growing row will be, but since the plants will not be allowed to grow too big in their nursery, there is no point in double-digging it. Few gardeners will need more than eight feet (240 centimeters) of nursery row since that much will produce 24 to 36 husky seedlings. After digging, rake the nursery smooth, sprinkle the fine seed thinly in narrow bands, half an inch (1.25 centimeters) wide, atop the earth. The bands should be about 12 inches (30 centimeters) apart. Then cover the seeds with fine soil or fine compost about a quarter inch (six millimeters) thick. With your hand, pat it down firmly enough to restore capillarity.

In maritime, short-season, and moderate climates, the time to start these seedlings is about when the earliest apple trees are blooming, before it gets too hot and while the soil will stay moist for a day or so after you lightly water it. This is when your seeds will germinate easily. Celery often dies in extremely hot weather, so where summers are steamy, consider it an autumn/winter crop and sow the seeds after the heat of summer breaks. To get the fine seed to germinate in this situation, you may have to erect shade over the nursery bed and water it frequently. Celery seed also germinates rather slowly, commonly taking 11 to 14 days to emerge.

Once the seedlings are up, thin them progressively so that when they are two to three inches (five to seven centimeters) tall they stand four to six inches (10 to 15 centimeters) apart. Keep the nursery bed well watered. Always. I locate my celery bed conveniently close to the garden faucet, and every time I use a hose, I direct it briefly at the celery bed. Nineteenth-century growers used to cut seedlings back an inch, like pruning a hedge, after they had grown to over four inches tall. Sometimes they would do this twice. This is a good idea because it makes the seedlings develop stronger roots that better withstand transplanting shock. After pruning the seedlings and growing them on, dig them up and transplant them into their growing row (which has been prepared as I suggested above.) When transplanting, take care to lift the seedlings with soil and roots intact, doing as little damage to the root structure as possible. Press the earth firmly around them after setting them in. You will have had time to grow and harvest an early crop on the row or bed that will receive the celery transplants; I suggest early peas. If growth slows after transplanting, side-dress the plants with some COF or chicken manure compost, or fertigate/foliar feed them.

Where water is short, or if you're mainly growing your garden on rainfall, I suggest direct-seeding small clusters of celery or celeriac seeds in their final growing positions and thinning each clump progressively down to a single one because transplanting destroys the taproot and converts the plant into more of a surface feeder. Direct-seeded crops endure rainless spells better.

Pests and diseases. Insects rarely trouble this crop when it is growing healthily. However, there are diseases. Most prevention involves rotation, not crowding the bed, and making sure the plants can continue growing rapidly. This is determined by soil conditions, fertility, and moisture; most of these conditions had to be created before the seeds were sown or transplants set out.

Varieties. Quality seed companies inevitably offer quality market varieties, but be cautious about betting the ranch on heirlooms. A lot of breeding work has gone into modern commercial varieties, such as Utah and Ventura, to prevent celery from bolting to seed in the first year and to build in resistance to various diseases. Be especially wary of cheap celeriac seed; it will not make much for you to eat after you have peeled away all the rough knobs and protuberances.

Harvest and storage. Where celery can withstand the frosts of winter, your harvest goes on until spring; in this case, snap off or cut off individual outer stalks as they are needed, or pull celeriac plants as wanted while the others continue sizing up slowly. If you are growing celeriac in a climate where winter brings only a short period of freezing cold weather, you might, as a precaution, want to hill soil up against, and an inch (2.5 centimeters) above, the edible balls, leaving most of their leafstalks exposed.

Where winter means snow and ice, your celery and celeriac crop should have become large enough to begin harvesting toward the end of summer. For winter use, celeriac is eminently cellarable and so is celery. Keep whole plants, roots and all, at close to freezing and they can overwinter. Celery can also overwinter in its growing bed if you have hilled earth up against it and have taken steps to prevent that soil from freezing so solid that you can't get at the plants when you want one. I suggest a careful reading of the Bubels' book on root cellars, or enjoy an interesting read of Henderson's classic *Gardening For Profit.*

Saving seed. You can grow seed if the plants overwinter. The flowering and maturation process is much like that for carrots, but celery will make thousands of small seed-bearing flowers instead of only a few per plant as carrots do. If a celery plant makes seed during its first year, and if that seed somehow matures, definitely do not save that seed for sowing (though it might be useful in the kitchen as seasoning). Celery is a biennial, and plants that act as annuals, going to seed quickly, are highly undesirable.

You should not try to maintain a fancy celery or celeriac variety unless you can manage a plant population of at least 25 in your gene pool. I urge seed savers who want to have a celery in their collection to grow one of the herbal types. These are much closer to wild celery and much more tolerant of less-than-ideal conditions.

Bibliography

There is no way this small book could provide all the information you need if you want to make vegetable gardening a serious business for your family's well-being and economic survival. I hope you will go on to read all the books listed below, many of the books listed in the bibliographies of those books, and so on.

Books

Albrecht, William A. *The Albrecht Papers.* Acres, U.S.A., 1976. Four volumes. The first two volumes are the most valuable. In print. The Soil and Health Library (see under "Online Resources" below) also provides a large collection of William Albrecht's writings that is just about as useful as the Acres collection.

Ashworth, Suzanne. *Seed to Seed: Seed Saving Techniques for the Vegetable Gardener.* Seed Savers Exchange, 1991. As the title says, the basic "how-to" of saving seed, vegetable by vegetable.

Bubel, Mike and Nancy. *Root Cellaring: The Simple No-Processing Way to Store Fruits and Vegetables.* Rodale Press, 1979. In print.

Coleman, Eliot. *The Four-Season Harvest: Organic Vegetables from Your Home Garden All Year Long.* Chelsea Green, 1992. In print. Coleman lives in Maine, where winter is wintery and then some, and grows winter salads and other items in the snow. Need I say more?

Deppe, Carol. *Breed Your Own Vegetable Varieties: The Gardener's and Farmer's Guide to Plant Breeding and Seed Saving.* Chelsea Green, 1993 and 2000. Whenever a gardener sets out to save their own seed, they *are* acting as a plant breeder. If this is done unknowingly, improperly, the variety you're saving can be ruined. Besides, breeding new varieties is fun. Carol's book is not hard to grasp, and makes an easy time of plant genetics.

Dufour, Rex. "Farmscaping to Enhance Biological Control" [online]. A pest management systems guide from ATTRA, December 2000. <www.attra.org/attra-pub/PDF/farmscaping.pdf>

Faulkner, Edward H. *Plowman's Folly*. Grosset & Dunlap, 1943. May be read online at the Soil and Health Library.

Henderson, Peter. *Gardening for Profit: A Guide to the Successful Cultivation of the Market and Family Garden*. Orange Judd Company, 1882. May be read online at the Soil and Health Library.

Hills, Lawrence. *Russian Comfrey: A Hundred Tons an Acre of Stock Feed or Compost for Farm, Garden or Small-holding*. Faber & Faber, 1953. May be read online at the Soil and Health Library.

Howard, Sir Albert, and Yeshwant D. Wad. *The Waste Products of Agriculture: Their Utilisation as Humus*. Oxford University Press, 1931. The best book ever done on composting. May be read online at the Soil and Health Library.

Howell, Dr. Edward. *Enzyme Nutrition: The Food Enzyme Concept*. Avery, 1985. Understand why to become a vegetableatarian. An earlier version, now out of print, may be read online at the Soil and Health Library.

Jeavons, John. *How to Grow More Vegetables Than You Ever Thought Possible on Less Land Than You Can Imagine*. Ten Speed Press, numerous editions between 1974 and 2002. Half a million copies have been sold; I consider this book misleading. In print.

Jenkins, Joseph. *The Humanure Handbook*. Jenkins Publishing, 1999. Available in print through the usual sources of retail trade. Because Jenkins wrote this book primarily as an act of service, it is also available for free download from many websites.

Krasil'nikov, N.S. *Soil Microorganisms and Higher Plants*. Academy of Sciences of the USSR (Moscow), 1958. Translated in Israel by Dr. Y. Halperin. There is no better introduction to the complexities going on under your feet. Quite readable. May be read online at the Soil and Health Library.

Koepf, H.H., B.D. Petterson, and W. Schumann. *Bio-Dynamic Agriculture: An Introduction.* Anthrosophic Press, 1976. The second-best book ever written on composting. May be read online at the Soil and Health Library.

Magdoff, Fred, and Harold van Es. *Building Soils for Better Crops.* 2nd ed., Sustainable Agriculture Network, 2002. As of 2005, available for free download at <www.sare.org/publications/index.htm/>.

Managing Cover Crops Profitably. A cooperative writing effort organized by the Sustainable Agriculture Network, USDA Agricultural Library, 1998. As of 2005, available for free download at <www.sare.org/publications/index.htm/>.

Oliver, George Sheffield. *Friend Earthworm: Practical Application of a Lifetime Study of Habits of the Most Important Animal in the World.* Oliver's Earthworm Farm School, 1941. May be read online at the Soil and Health Library.

Pfeiffer, Ehrenfried. *Soil Fertility, Renewal and Preservation.* Faber & Faber, 1947. May be read online at the Soil and Health Library.

Poisson, Leandre and Gretchen. *Solar Gardening: Growing Vegetables Year-Round the American Intensive Way.* Chelsea Green, 1994. I do not agree with their ill-considered intensive spacing recommendations or their derogation of extensive systems as "non-productive," but this book does introduce many interesting structure-based season extenders that are useful where the snow flies in winter.

Solomon, Steve. *Growing Vegetables West of the Cascades.* 5th ed., Sasquatch, 2000. In print. The fourth edition (1988) is also quite good. I do not recommend my first three editions.

"Vegetables Without Vitamins" [online]. *Life Extension Magazine;* March 2001. <www.lef.org/magazine/mag2001/mar2001_report_vegetables.html>.

Weaver, John E., and William Bruner. *Root Development of Vegetable Crops.* McGraw Hill, 1927. May be read online at the Soil and Health Library.

Online resources

ATTRA. Appropriate Technology Transfer for Rural Areas is the national sustainable agriculture research service funded by the USDA. Good folks here. <www.attra.ncat.org/>

Cooperative Extension Services. Those from the states of North Carolina <www.ces.ncsu.edu/>, Louisiana <www.lsuagcenter.com/>, and New York <www.cce.cornell.edu/> provide the largest range of the most carefully considered online materials as of 2005.

Soil and Health Library. A free online public library offering a collection of classic holistic farming and gardening books and articles, most issued prior to 1960. The serious student of holistic farming or gardening would be wise to read every title in this online resource. <www.soilandhealth.org/>

Sustainable Agriculture Research and Education. Offers a substantial collection of topical books for free download. <www.sare.org/>

Index

About the Author

I started out growing a large vegetable garden in the San Fernando Valley, California, 1973. During the early years I attempted all the usual homesteader foibles—chickens, rabbits, fattening a steer. I even tried market gardening using biodynamic French intensive methods, and failed miserably. I moved to an Oregon 5 acre homestead in 1978. After searching for a legal cash crop, I decided to raise the consciousness of fellow veggie gardeners, and started Territorial Seed Company and wrote *Growing Vegetables West of the Cascades*. This opened the door to what has been my life's work—encouraging at-home food self sufficiency. By 1985 the business had turned into a monster; my book was in its third edition. I sold Territorial and devoted myself thereafter to enjoying independent poverty and a homestead-based lifestyle. I moved to Canada in 1994, to Australia in 1998. I am still writing veggie gardening books, giving classes and doing everything I can to help people become financially independent through producing their own necessities. I am also proud to be the creator of an online library, soilandhealth.org, offering key books on the connections between soil fertility and human health.

If you have enjoyed *Gardening When It Counts*
you might also enjoy other

BOOKS TO BUILD A NEW SOCIETY

Our books provide positive solutions for people who want to make a difference. We specialize in:

Environment and Justice • Conscientious Commerce
Sustainable Living • Ecological Design and Planning
Natural Building & Appropriate Technology • New Forestry
Educational and Parenting Resources • Nonviolence
Progressive Leadership • Resistance and Community

New Society Publishers

ENVIRONMENTAL BENEFITS STATEMENT

New Society Publishers has chosen to produce this book on Enviro 100, recycled paper made with **100% post consumer waste**, processed chlorine free, and old growth free.

For every 5,000 books printed, New Society saves the following resources:[1]

37	Trees
3,315	Pounds of Solid Waste
3,648	Gallons of Water
4,758	Kilowatt Hours of Electricity
6,027	Pounds of Greenhouse Gases
26	Pounds of HAPs, VOCs, and AOX Combined
9	Cubic Yards of Landfill Space

[1] Environmental benefits are calculated based on research done by the Environmental Defense Fund and other members of the Paper Task Force who study the environmental impacts of the paper industry.

For more information on this environmental benefits statement, or to inquire about environmentally friendly papers, please contact New Leaf Paper – info@newleafpaper.com Tel: 888 • 989 • 5323.

NEW SOCIETY PUBLISHERS